QUANTITATIVE MINERAL RESOURCE ASSESSMENTS

QUANTITATIVE MINERAL RESOURCE ASSESSMENTS

An Integrated Approach

Donald A. Singer

W. David Menzie

OXFORD
UNIVERSITY PRESS
2010

OXFORD
UNIVERSITY PRESS

Oxford University Press, Inc., publishes works that further
Oxford University's objective of excellence
in research, scholarship, and education.

Oxford New York
Auckland Cape Town Dar es Salaam Hong Kong Karachi
Kuala Lumpur Madrid Melbourne Mexico City Nairobi
New Delhi Shanghai Taipei Toronto

With offices in
Argentina Austria Brazil Chile Czech Republic France Greece
Guatemala Hungary Italy Japan Poland Portugal Singapore
South Korea Switzerland Thailand Turkey Ukraine Vietnam

Published by Oxford University Press, Inc.
198 Madison Avenue, New York, New York 10016

www.oup.com

Library of Congress Cataloging-in-Publication Data
Singer, D. A.
Quantitative mineral resource assessments : an integrated approach /
Donald A. Singer, W. David Menzie.
p. cm.
Includes bibliographical references and index.
ISBN 978-0-19-539959-2
1. Mines and mineral resources.
2. Geology, Economic. I. Menzie, W. D. II. Title.
TN153.S553 2010
553—dc22 2009015157

To
John Cedric Griffiths

Preface

This book is the result of almost forty years of concentrated work by the senior author. Many of the basic concepts presented in the book were collaboratively developed by the authors between 1977 and 1987, when they occupied adjacent offices at the U.S. Geological Survey in Menlo Park, California. Over the following twenty years, many of those concepts were further developed, elucidated, and exemplified in the professional literature. From the beginnings of the work, colleagues have asked for a unified presentation of the many ideas about quantitative resource assessments (QRA) that are located in widely disperse publications. Although not apparent from individual reports, there was a broad plan for the kind of QRA presented here from the beginnings of this work in 1974–1979. The purpose of the form of mineral resource assessments discussed in this book is to make unbiased quantitative assessments in a format needed in decision-support systems so that consequences of alternative courses of action can be examined with respect to land use or mineral-resource development. The audiences for these assessments are governmental or industrial policy-makers, managers of exploration, planners of regional development, and similar decision-makers. Some of the tools and models presented here are useful for selection of exploration sites, but that is a side benefit, not the goal.

Readers of this book are likely to be either users or practitioners of assessments. For them, we brought together materials published in diverse places and have tried to capture the necessary ingredients of the diverse disciplines that are an integral part of quantitative mineral resource assessments. We believe that the book is written so that the procedures are relatively easy to understand for those who come from one of the diverse disciplines. The cost of writing for the broader audience is that the mineral economist who reads the book might understand that knowing the geologic setting is important but may not understand the complexity involved in determining the setting. The same can be said of the economic geologist, who may not understand the implications and assumptions made when a Poisson distribution is assumed for the distribution of deposits or when a net present value is used to determine worth. The point is that although the concepts presented here are relatively straightforward and understandable to many, in assessments, carefully listening to the experts in other disciplines leads to better products. Borrowing from De Veaux and Velleman's (2008) comments about the challenges of teaching statistics, we believe that navigating through and making sense of QRA require not just learning rules and equations but also life experiences and common sense. The judgment required to understand which tools to apply is best learned by example and experience.

The accomplishments reported here reflect many years of work by a group of dedicated researchers. Their publications do not convey the difficulties of their work. Those who made significant contributions include Dan Mosier, Jim Bliss, and Greta Orris. Important additional contributions were made by Dave Root, Norm Page, Keith Long, Vladimir I. Berger, and Joe Duval. The suggestions and critiques over many years by Barney Berger, Larry Drew, Margie Scott, and Ryoichi Kouda have led to many improvements and are greatly appreciated. Over the years, reviews, questions, and suggestions on papers made by Paul Barton, Brian Skinner, Roderick Eggert, and Don Sangster have helped improve these works. Our common training in mineral economics from DeVerle P. Harris at Penn State University provided a basis for decision support systems.

Although he would probably deny any responsibility, none of this would have been possible without the training that we received from the remarkable John Cedric Griffiths at Penn State University. He taught us how to focus on the ultimate goal and not the items or tools.

Finally, we would like to thank Brian Singer at altitudesf.com for providing the cover art.

Contents

QUANTITATIVE MINERAL RESOURCE ASSESSMENTS

1

Introduction

Perspective

Every day, somewhere in the world, decisions are made about how public lands that might contain undiscovered resources should be used or whether to invest in exploration for minerals. Less frequently, decisions are made concerning mineral resource adequacy, national policy, and regional development. Naturally, the people making the decisions would like to know the exact consequences of the decisions before the decisions are made. Unfortunately, it is not possible to inform these decision-makers, with any certainty, about amounts, discoverability, or economics of undiscovered mineral resources. The kind of assessment recommended in this book is founded in decision analysis in order to provide a normative framework for making decisions concerning mineral resources under conditions of uncertainty. Our goal is to make explicit the factors that can affect a mineral-related decision so that the decision-maker can clearly see the possible consequences of the decision. This means that we start with the question of what kinds of issues decision-makers are trying to resolve and what types and forms of information would aid in resolving these issues. This book has a different purpose than academic reports common to many assessments, and it is not designed to help select sites for exploration. The audience for products of assessments discussed here comprises governmental and industrial policy-makers, managers of exploration, planners of regional development, and similar decision-makers. Some of the tools and models presented here are useful for selection of exploration sites, but that is a side benefit. The focus of this book is on the practical integration of the fundamental kinds of information needed by the decision-maker. The integrated approach to assessment presented in this book focuses on three assessment parts and the models that support them. The *first part* uses models of tonnages and grades to estimate possible tonnages and grades of undiscovered deposits. The *second part* develops mineral resource maps that explore whether an area's geology permits the existence of one or more types of mineral deposits. The product of this part of the assessment is identification of so-called permissive tracts of land. For those areas that are permissive, the *third part* of the assessment develops estimates of the possible number of undiscovered deposits of different types. These estimated undiscovered deposits are consistent with the grade-and-tonnage models of the first part. These three parts are the centerpiece of this book; the mineral deposit models are used to aid in construction of the three parts and to help convert the output of the three parts into forms helpful to decision-makers.

We believe that *qualitative* assessments of undiscovered minerals are most useful when there is shared understanding and trust between the assessor and the decision-maker, for example, in a small company where the decision-maker is comfortable with the advice of the geologist. In most situations, however, this is not the case because organizations are too large or there are too many interested parties. When three or more parties have an interest in the outcome of the decision, there may be a need to adjudicate. Qualitative assessments are typically subjective and so poorly defined that implications of such statements as "high potential for X" cannot be documented and defended in an adversarial situation. So, in most cases, some form of *quantitative* assessment is warranted.

Many papers and books dealing with academic issues of quantitative mineral resource assessment are interesting; we recommend them to stimulate thought and discussion. Maurice Allais is considered the father of modern quantitative resource assessment not because he was the first to use quantitative methods (see King, 1880), but because he was probably the first to design a system to respond to the needs of decision-makers (see Allais, 1957). The next significant steps in the development of quantitative mineral resource assessments were proposed by Harris (1965, 1984). He extended the work of Allais by introducing multivariate methods, using economics, capturing the knowledge of geologists, and focusing on the decision-maker. This book has the same goal as did Allais and Harris—to provide information useful to decision-makers. The book represents the work of a dedicated team of scientists which, in a span of thirty years, constructed and refined numerous models, tested methods, and participated in assessments.

The differences between the ideas presented by Allais fifty years ago and Harris twenty to forty years ago, and those presented in this book reflect a significant growth in knowledge gained since their works and the recognition of the value of, and improved ways to capture, geologic information.

Owing to substantial advances in understanding about the earth's nature, we now can use geologic maps to divide a large region into parts that could contain different kinds of mineral deposits, and we know that these different kinds differ significantly in the amounts and qualities of minerals of interest to society, which affects chances that the deposits will be sought, found, and exploited by society. These topics are discussed below in the context of a brief history of three-part assessments and a guide to the remaining parts of this book. It is important to recognize that selections of methods presented here are motivated by the desire first to make unbiased estimates and then to minimize the uncertainty associated with the estimates.

Beginnings of Three-Part Assessments

In the not too distant past, the directors of a major mining company were inclined to invest in the new larger trucks for the company's existing mine due to the almost certain 15 percent return on the investment, rather than in the vague hopes of large financial returns that might occur at some point in the future if the monies were invested in mineral exploration. Heads of exploration have found themselves in this situation many times over the years because boards must meet their fiduciary responsibilities and be concerned about risk. How could the vice president of exploration compete for exploration funds? At a minimum, the board required an estimate of the probability of success, an estimate of the value of the resources being sought, and knowledge of exploration costs. He could neither specify the expected monetary reward nor the uncertainty of his proposed exploration investment.

Over the last forty years, some companies have responded to this competition for limited funds by conducting internal studies of the possible number of undiscovered deposits in a region, the possible values of these deposits, and the chances of finding them. Classifying the deposits into types considerably reduces the work of determining their possible values, the geologic settings where the deposits might occur, and the chances and costs of discovery. Quantification of this information provides a foundation for analyzing the uncertainty and risk of the exploration investment. Thus have some exploration programs been justified and funded. For the most part, these studies have been considered company confidential and are not publicly available. However, the key elements of these studies are the same as those needed for decisions about public lands, regional development, and resource development issues that are discussed in a series of examples of histories of actual assessments given in boxes throughout this book. A key to the development of modern resource assessments was recognition that differences in locations, amounts, and qualities can be captured by knowing about different kinds of mineral deposits that contain the resources of interest. From the standpoint of the field of statistics, the variability of locations and values of mineral resources are partitioned by deposit types. For example, the total variability in amounts of some metal can be explained and predicted by knowing whether a particular deposit type could or could not occur in an area of interest. Thus, the large uncertainties inherent in estimating amounts of undiscovered resources in an area of interest are significantly reduced by knowing if a deposit type could occur. The same could be said for the values and possible locations of these undiscovered resources.

Mineral deposit models are important in quantitative resource assessments for two reasons: (1) numbers per unit area of deposits and grades and tonnages of most deposit types are significantly different, and (2) types occur in different geologic settings that can be identified from geologic maps. As a consequence, mineral deposit models can be used to reduce uncertainty about locations, number of deposits, amounts of resources, and values of the resources. Mineral deposit models are the keystone in combining the diverse geoscience information on geology, mineral occurrences, geophysics, and geochemistry used in resource assessments and mineral exploration.

Perhaps the first publications showing the power of delineating tracts of land for deposit types using geology and using grade-and-tonnage models with associated estimates of the number of undiscovered deposits was in an assessment of a 1:250,000-scale quadrangle in Alaska (Richter, Singer, and Cox, 1975; Singer, 1975). An order of magnitude increase in complexity came in a 1:1,000,000-scale assessment of about 80 percent of Alaska in which eleven new grade-and-tonnage models were developed and 144 tracts were delineated with eighty-five estimates of number of deposits (Eberlein and Menzie, 1978; Grybeck and DeYoung, 1978; Hudson and DeYoung, 1978; MacKevett, Singer, and Holloway, 1978; Singer and Ovenshine, 1979).

ASSESSMENT EXAMPLE

As part of the 1958 Alaska Statehood Act, the state was entitled to select 102 million of the state's 375 million acres (1.520 million km^2) for whatever use the state decided. Additionally, Alaskan Natives were entitled to select 44 million acres for ownership as a result of the 1971 Alaska Native Claims Settlement Act, and in another Act, the U.S. Interior Department had rights to withdraw up to 80 million acres in several land categories. The U.S. Congress was required to complete these various classifications by 18 December 1978. Each affected party had an interest in and desire for information about discovered and possible undiscovered mineral resources in Alaska. Responding to this need, the U.S. Geological Survey (USGS) in 1977 began a 1:1,000,000-scale, three-part quantitative mineral resource assessment of about 80 percent of the state. Without knowing decision-makers' intents before the assessment was released in February 1978, it is not possible to determine how the assessment changed their decisions. We do know that the State of Alaska selected land now known to contain one of the largest copper deposits in the world (Pebble Copper) located in a tract delineated in the assessment as permissive for porphyry copper deposits but containing no known deposits in 1978, and which had estimates of undiscovered copper deposits. We also know that in central Alaska some important gold deposits were discovered after the assessment that were not recognized as a known type at the time of the assessment. One of the major zinc-lead concentrations in the world was recognized in the Brooks Range of Alaska by the assessment, but its boundaries were not well defined due to sparsely known geology in that part of Alaska.

Many of the economic geologists, statisticians, mineral economists, and computer specialists who participated in the 1:1,000,000-scale assessment of Alaska became the core group that continued the development and applications of what was later called the three-part form of resource assessment. An assessment of Colombia's mineral resources led to the first publication of descriptive (Cox, 1983a, 1983b) and linked grade-and-tonnage models (Singer and Mosier, 1983a, 1983b) that were precursors to a more compressive compilation of mineral deposit models authored by thirty-seven scientists (Cox and Singer, 1986). A robust method of combining probabilistic estimates of number of deposits, grades, and tonnages without relying on some strong assumptions was developed by David Root (see Root, Menzie, and Scott, 1992).

The methods, procedures, and models developed in support of this form of assessment were finally given the name "three-part assessments" in 1993 (Singer, 1993a). The word "part" is used rather than "step" to indicate that these assessments are not always done in the same sequence. More important than the name was the recognition that the success of this form of assessment depends on it being an integrated approach. By this we mean that no part of this system of models and methods of estimation has any meaning in isolation. For instance, estimates of the number of undiscovered deposits are completely arbitrary unless tied to a grade-and-tonnage model. As demonstrated throughout this book, the various models and methods are linked to the other parts. This integration of models and methods is a source of strength that reduces the chances of biased estimates, but it is also a burden, requiring careful development of the kinds of models and applications of methods discussed in this book.

Chapters of this book are divided into four groups. Chapters 1 and 2 provide information about why assessments are done and the nature of mineral resources. Mineral deposit models covered in chapters 3 through 5 are key to three-part assessments. Details concerning the three parts of the recommended form of assessment are presented in chapters 6 through 8. Chapters 9 through 11 focus on ways to use the assessment parts.

Important Ideas on Assessments

Because the primary purpose of the kind of assessments recommended here is not to locate where to drill or mine but to help decision-makers determine consequences of economic and policy decisions about tracts of land, regions, or countries, it is critical that the assessments be unbiased. Technically, a statistic is biased if, in the long run, it consistently overestimates or underestimates the parameter it is estimating. This idea can be visualized by considering a target and three different attempts to hit its center (figure 1.1). The average center of a shooter's attempts can be compared with the center of the target: when the two are different, we can say that the shooter is not accurate or is biased. In assessments of undiscovered mineral resources, we do not want biased estimates. The problem is that we typically cannot see the target (undiscovered resources) in these assessments. However, we can do some things to reduce the chances of introducing bias in our estimates. For example, when someone tries to make a "conservative" estimate to reduce the chances of being wrong, or makes a high estimate to allow for the unknown, that person is introducing a bias of some unknown amount, and this should be resisted. Examples of situations that can introduce biases or examples of how certain parts of assessments address biased estimates are presented in many chapters of this book. In addition, the integrated approach to assessments recommended in this book is specifically designed to reduce chances

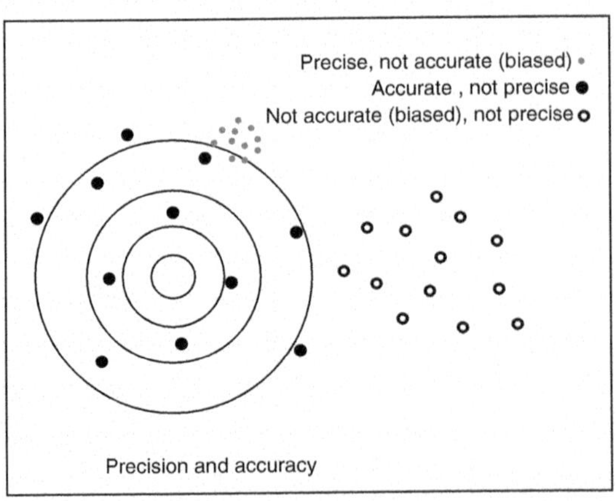

Figure 1.1 Target and three different attempts to hit its center. Variation about the center of each cluster represents precision, whereas the difference between the center of each cluster and the target center represents accuracy or bias.

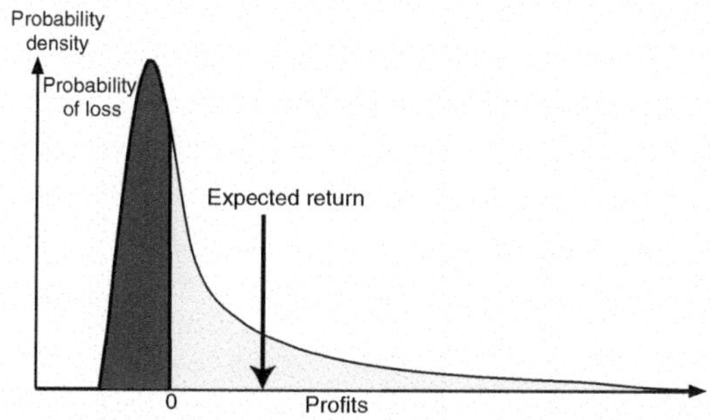

Figure 1.2 Hypothetical probability of return from investment and expected return. In exploration, it is possible to both have a significant positive expected profit and also have a significant chance of losing money. Awareness of this should affect decision-making.

of biased estimates. Before introducing the various parts and models used in these assessments, it is helpful to discuss some aspects of mineral resources that influence how and why assessments should be performed.

Chapter 2 provides an overview on the supply and demand of mineral resources as well as a perspective on why certain aspects of resources are of greater importance than others and why certain kinds of mineral deposits are of greater interest than others. Demand and supply of minerals follow patterns of regional development that affect prices and have implications for assessments. In particular, we show that only certain deposit types have dominated supply. In addition, only the largest deposits are important to global supply. The effects of these distributions of sizes and types of deposits are the greatest sources of uncertainty in assessments and also affect risk of exploration failure. It is the frequency distribution of deposit sizes that caused Allais (1957) to observe for his study that the expected financial return from an exploration investment was positive, but the probability of economic failure was 0.65 (figure 1.2). Deposit models help to reduce both uncertainty and risk of failure.

Mineral Deposit Models for Three-Part Assessments

A rough overview of the role of the various mineral deposit models in three-part assessments is shown in figure 1.3. Mineral deposit models are the focus of chapters 3 through 5. The models are the keystone in combining the diverse geoscience information on geology, mineral occurrences, geophysics, and geochemistry used in resource assessments and mineral exploration. Although these different kinds of models are presented separately in this book, it should not be forgotten that they do not exist in isolation. None of these models has any useful meaning except in the context of the other related models of the same deposit type. Construction of these models is an iterative process where changes in a deposit type's descriptive model, for example, affects and necessitates changes in the type's grade-and-tonnage model. It is the internal consistency of these models that provides the strong foundation upon which unbiased assessments can be made—achieving this internal consistency requires much time-consuming work for their proper construction. Descriptive mineral deposit models (chapter 3) are critical to construction of the first part of assessments, the grade-and-tonnage models (chapter 6). Additionally, descriptive models are the primary source of guidance for linking geoscience information to deposit types in the second part of assessments, delineation of permissive tracts (chapter 7). In order to properly serve this function, descriptive models focus on observations, use theories of deposit origins only to suggest what to observe, and should have their model properties documented at the scale of assessments.

In estimating the number of mineral deposits, a robust method is based on mineral deposit densities (chapter 4), which are a form of mineral deposit model wherein the numbers of deposits per unit area from well-explored

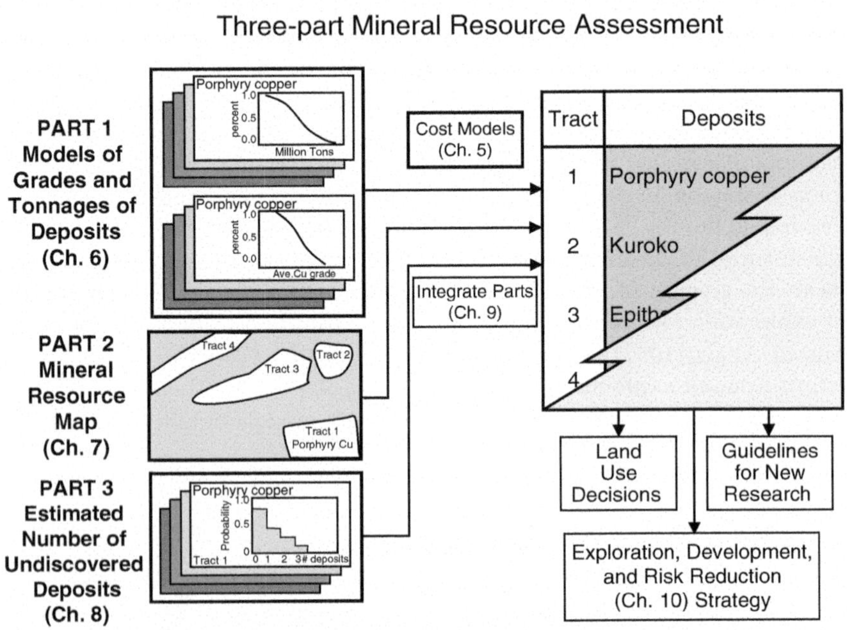

Figure 1.3 Relationships of book chapters to three-part assessments. Chapters 1 and 2 provide overviews of resources and assessments. Descriptive models (not shown) (chapter 3) guide part 2, the mineral resource map (chapter 7). Grade-and-tonnage models (chapter 6) serve as part 1 and assist density models (not shown) (chapter 4) in guiding part 3, estimating number of deposits (chapter 8). Cost models (chapter 5) aid in integrating parts 1 through 3 (chapter 9).

regions are counted and the resulting frequency distribution enables estimation of the number of undiscovered deposits (chapter 8). Deposit densities can be used either directly for an estimate or indirectly as a guideline in some other method.

In resource assessments of undiscovered mineral deposits and in the early stages of exploration, including planning, a need exists for prefeasibility cost models (chapter 5). These models, which separate economic from uneconomic deposits, help assessors and decision-makers focus on targets that can benefit society or the exploration enterprise. In three-part assessments, these models can be used to eliminate deposits that would probably be uneconomic even if discovered. Awareness of economics of deposit sizes has helped determine priorities of deposit types selected for constructing grade-and-tonnage models.

ASSESSMENT EXAMPLE

The U.S. Forest Service asked the USGS and the U.S. Bureau of Mines to quantitatively assess the future copper and silver production possibilities from sediment-hosted copper deposits within the Kootenai National Forest of Idaho and Montana. The three-part assessment provided estimates of the number of undiscovered deposits, associated grade-and-tonnage distributions, and an economic analysis of the resources (Spanski, 1992). Mineral deposit models and estimates of the number of deposits were integrated with mine and mill cost-estimation models and relevant economic and policy assumptions to estimate possible mineral production and associated economic effects (Gunther, 1992), including number of jobs generated and impacts of land-use policies.

Three Parts of an Assessment

Grade-and-tonnage models occupy a position of being both a kind of deposit model and also one of the three parts of assessments. Frequency distributions of tonnages and average grades of well-explored deposits of each type are employed as models for grades and tonnages of undiscovered deposits of the same type in geologically similar settings (chapter 6). Data utilized to construct these models include average grades of each metal or mineral commodity of possible economic interest and the associated tonnage based on the total production, reserves, and resources at the lowest possible cutoff grade. These data represent an estimate of the endowment of each of many known deposits so that the final models can accurately represent the endowment of all undiscovered deposits.

ASSESSMENT EXAMPLE

In an attempt to provide information about undiscovered mineral resources to the various parties selecting lands in Alaska by 1978, a nonquantitative approach was prepared by a group of geologists in the USGS. The map indicated high potential for chromium across a broad swath in central Alaska. What was meant by "high potential" in this case was that a large number of undiscovered podiform chromite deposits were believed to exist in this zone. Fortunately, a quantitative assessment was proceeding at the same time by another group that had analyzed the distribution of sizes of podiform chromite deposits, which showed that even if large numbers of deposits existed, the sum of the chromite resource could not be large enough to affect national demand. About the same time, attorneys and economists were considering the U.S. government's position with respect to embargos of chromite from some countries in southern Africa. After the grade-and-tonnage model was shown to the USGS managers, the qualitative assessment was withdrawn in order to prevent the possibility of it being misinterpreted to indicate that the United States had large undiscovered chromite resources.

The geologic settings used to define the mineral grade-and-tonnage models and descriptive models are key to identifying where the deposit type could occur, which is the second part of an assessment. To be able to consistently assess the undiscovered mineral resources of regions, areas are delineated where geology permits the existence of deposits of one or more specified types. These areas, called permissive tracts, are based on geologic criteria derived from deposit models that are themselves based on studies of known deposits within and outside the study area (chapter 7). Tracts may or may not contain known deposits.

The third part of an assessment is the estimation of some fixed, but unknown, number of undiscovered deposits of each type that exist in the delineated tracts (chapter 8). Until the area being considered is thoroughly and extensively drilled, this fixed number of undiscovered deposits, which could be any number including 0, will not be known with certainty. Estimates of the number of deposits explicitly represent the probability (or degree of belief) that some fixed but unknown number of undiscovered deposits exist in the delineated tracts. As such, these estimates reflect both the uncertainty of what may exist and a measure of the favorableness of the existence of the deposit type. In many cases, experts make estimates of number of deposits, so chapter 8 also includes new evidence that such estimates can be unbiased and provides some robust guides to reduce chances of biased estimates.

Integrating the Three Parts and Including Economics

If the decision-makers needed information only on the general locations of undiscovered mineral resources, the output of part two, delineation of permissive tracts (see figure 1.3), would be adequate for these assessments. If assessments were conducted only to estimate amounts of undiscovered metals, we would need contained metal models and estimates of the number of undiscovered deposits. Grades are simply the ratio of contained metal to tons of ore, so contained metal estimates are available for each deposit. In the simplest of all cases, one could multiply the average contained metal in a deposit type by the expected number of undiscovered deposits to make an estimate of the expected metal endowment. Such an estimate would be of limited value to a decision-maker because it neglects the considerable uncertainty that should be attached to the estimate, and it ignores the issue of whether the metal might be economic to extract. Ways to capture this uncertainty and the economic effects without violating important assumptions about resources are discussed in chapter 9.

Questions about the economics of undiscovered mineral deposits are central to one of the original reasons for developing the three-part assessment. Although the objectives and aversion to risk may be different for mining enterprises and governments, the kind of information needed by the decision-makers is similar (Scott and Dimitrakopoulos, 2001). Reduction of risk of exploration failure is one of the principal values of geologic information provided by governmental organizations. The ability to quantify the uncertainty allows examination of the risks of exploration (chapter 10). The risks of financial loss in exploration stem in large part from the nature of mineral resources, as discussed in chapter 2. There are some ways to reduce these risks but not eliminate them. The keys are to recognize the extent of the risks in the early planning stages and to understand the trade-offs between costs of gathering additional information to reduce risk and the benefits of these reduced uncertainties.

In chapter 11, we point out where there are opportunities to improve the three-part form of mineral resource assessment. Many of these opportunities come from identified sources of uncertainties in assessments of all kinds, such as assessing resources under cover. Some of the improvements can be made in assessments that are not completed, such as including economic filters. Additional opportunities come from the possibilities of harnessing the power of new technologies such as probabilistic neural networks to well-designed applications in these kinds of assessments.

2

Mineral Resources and Society

Perspective

Modern society cannot live without electric and electronic products, concrete, glass, fertilizers, ceramics, motor vehicles, airplanes, refrigerators, stoves, and medical equipment, all of which are made with products of mining. In the 1950s and again in the 1970s there was serious concern about whether we would run out of mineral resources. This recurring theme is driven largely by the increasing amounts of mineral material produced from mines and used by society over time.

One of the most striking aspects of the increasing quantities of mineral materials produced has been that prices of many minerals have been declining for more than 100 years (e.g., figure 2.1). Historically, prices of nonfuel mineral materials have declined relative to consumer goods and wages (Barnett and Morse, 1963). The declining prices have had a positive influence on general economies of mineral users by reducing prices of the factors of production of finished goods. Because mineral commodities are the building blocks of so many industries and products, the declining prices reverberate throughout the economy. Declining mineral commodity prices have largely been due to the successes of mining engineers in repeatedly lowering mining and processing costs and of geologists in lowering discovery costs of mineral deposits. Demonstrating the variability of commodity prices, between 2003 and 2008 prices have dramatically increased,

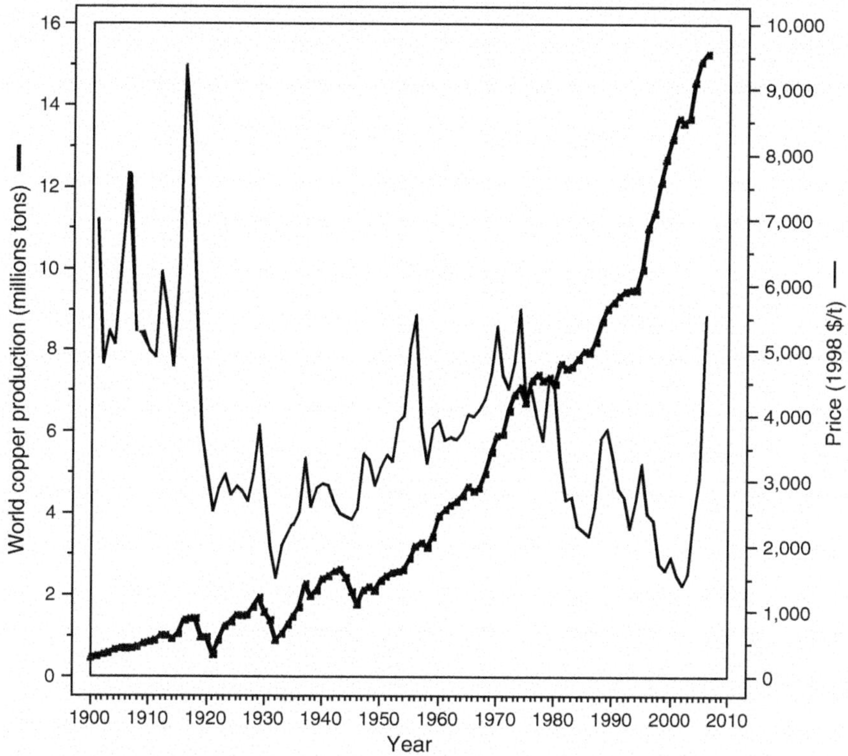

Figure 2.1 World copper production compared to price in 1998 U.S. dollars. Consumer Price Index used for conversion to 1998 U.S. dollars. Data from Porter and Edelstein (2008).

and in 2008 they declined again. Understanding how it is possible to have both increasing production and decreasing and more recently increasing and then decreasing prices of minerals is important to assessors and to decision-makers.

Decision-makers, whether concerned about regional development, exploration, or land management, are faced with the dilemma of obtaining new information, or allowing or encouraging others to obtain it, and the possible benefits and costs of development if mineral deposits of value are discovered. The intent in this chapter is to provide decision-makers and assessors a modern perspective on the geologic controls of mineral supply and demand and on the importance to supply of different kinds of mineral deposits and occurrences.

ASSESSMENT EXAMPLE

A quantitative mineral assessment was undertaken at the request of U.S. Department of State to stimulate investment in the Bolivian mining sector in the wake of the collapse of the tin market. The collapse left many miners without mining employment. The Department of State, which did not want to see the miners turn to growing cocoa, encouraged the U.S. Trade and Development Program (TDP) to fund the program. The TDP sought to provide the groundwork for further development of the Bolivian mining sector while at the same time introducing U.S. industry to business opportunities there. The joint assessment served as a vehicle to better inform the mineral exploration community and to reduce the exploration risk to mining companies (U.S. Geological Survey and Servicio Geologico de Bolivia, 1992).

Geologic Supply

Supply of minerals to society is dependent not only on the total amount of mineral material but also on the quality or concentrations, the spatial distributions or how scattered the material is, whether it has been found, whether it is remote from infrastructure, and a whole host of other issues such as government policies, production technologies, and market structures. These nongeologic determinants of mineral supply are not considered here. Most uses of metals do not reduce their amounts, but only relocate and reconstitute them, so why should anyone be interested in questions about the supply of minerals to society? An important reason is that we simply do not know very much about possible supplies of these materials and would prefer to have less uncertainty about factors of such importance to society. The earth contains vast quantities of every metal we use. Even if estimates of crustal abundances are an order of magnitude too high, there are still immense quantities of metals.

The question is how much metal is available in concentrations and forms of interest to society. As long as costs of extraction of metals are above zero, industry will seek out the concentrations and forms of metals that have cost-reducing physical attributes (DeYoung and Singer, 1981). Both form, or mineralogy, and concentration, or grade, are critical to consider because either can make some of the total amount of metal inaccessible due to excessive costs of extracting and processing the materials of interest. In addition, current concepts of cost go beyond traditional views of direct production costs to include, for example, chemical attributes that increase environmental costs, whether they are internalized or not.

Most of the earth's less common metals, such as copper, zinc, lead, and nickel, occur more or less evenly distributed in common silicate minerals

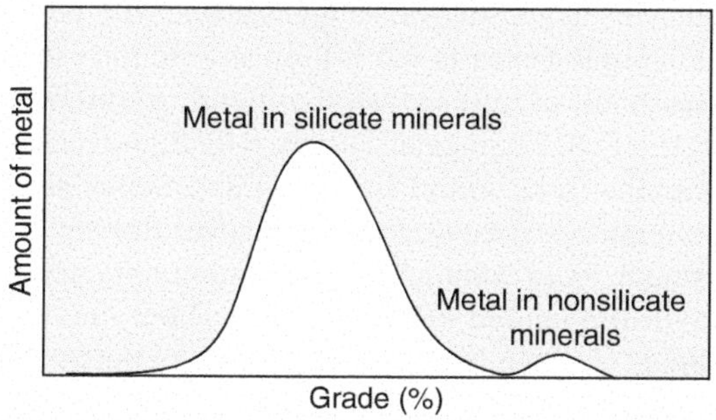

Figure 2.2 Probable distribution of geochemically scarce metals in the earth's crust. After Skinner (1976).

(Skinner, 1976). Metals in this form cannot be significantly concentrated to reduce volumes to be processed, and the entire mineral must be broken down chemically to separate the desired metal from the other atoms. Because the chemical bonds in most common minerals are so strong, this is a complicated and very energy-intensive process (Skinner, 1976). For these reasons, mining of these scarce metals focuses on rare mineral deposits that contain metal in compounds with elements such as sulfur or oxygen. The metals in such deposits are more easily extractable and are found in significant concentrations or grades (figure 2.2). Total amounts of copper, zinc, lead, silver, and gold in extractable minerals that are contained in all discovered mineral deposits range from 0.015 to 0.002 percent of the amounts contained in the upper 1 km of continental crust (Singer, 1995). Thus, presently identified extractable minerals in deposits represent a very small percentage of the total amount of metal.

To demonstrate that only a small percent of the metallic elements in the earth's crust are contained in identified mineral deposits, several researchers have constructed plots of the metals in mineral deposits or ore deposits versus metals in rocks (usually the upper 1 km of the earth's crust) (McKelvey, 1960, 1973; Erickson, 1973; Brooks, 1976; Barton, 1983). Barton (1983, p. 6) pointed out that the large amounts of metals that are not accounted for in identified conventional metal deposits offer great opportunities for undiscovered deposits of these metals. These and similar analyses generated renewed speculation about the promise held by the observed relation between decreasing grades acceptable for mining and resulting increases in mineral availability (Lasky, 1950). However, the promise is not kept because the mathematical model used as a basis for such extrapolations has been demonstrated to produce physically impossible results outside of the limited range of observed mineral-deposit grades (DeYoung, 1981).

It is quite difficult to estimate the total amounts of the earth's metals that are in undiscovered mineral deposits that might become available for mining. Based on an extrapolation of copper in one of the world's most heavily mineralized areas in the southwestern United States, the National Academy of Sciences (COMRATE, 1975) estimated that the upper limit on the quantity of copper in ores with grades greater than 0.1 percent that can be produced in the United States is 0.9 billion metric tons (Gt). In a detailed probabilistic assessment of undiscovered mineral resources using the three-part assessment, the U.S. Geological Survey (USGS) estimated that about as much copper remains to be found in the United States as has been found to date (Ludington and Cox, 1996; U.S. Geological Survey National Mineral Resource Assessment Team, 1998). The total amount of copper discovered in mineral deposits in the United States is 0.4 Gt; about one-fourth of this has already been mined. These independent estimates of the total amount of copper in mineral deposits in the United States are remarkably consistent—an upper limit estimate of 0.9 Gt versus a total of resources in known deposits plus undiscovered deposits estimate of 0.8 Gt. Thus, assuming that it is at acceptable grades, deposit sizes, and discoverable, about seven-eighths of the total resources (identified and undiscovered) of copper in the desirable sulfide or oxide form in the United States should yet be available for mining.

But are these resources in deposits with acceptable grades, sizes, and locations? Worldwide, at least 62 percent of the 260,000 metric tons of gold discovered to date is located in four countries, and more than 68 percent occurs in four types of mineral deposits. About 55 percent of the 2,400,000 metric tons of silver found is in four countries, and 45 percent is in four types of deposits. Fifty-six percent of the 2.1 billion metric tons of discovered copper is from four countries, and three types of deposits contain 86 percent of the total (figure 2.3). More than 50 percent of both the 890,000,000 metric tons of zinc and 460,000,000 metric tons of lead discovered to date come from four countries, and 70 percent of both metals occur in four types of deposits. All discovered gold would fit in a cube with a height of 24 m,

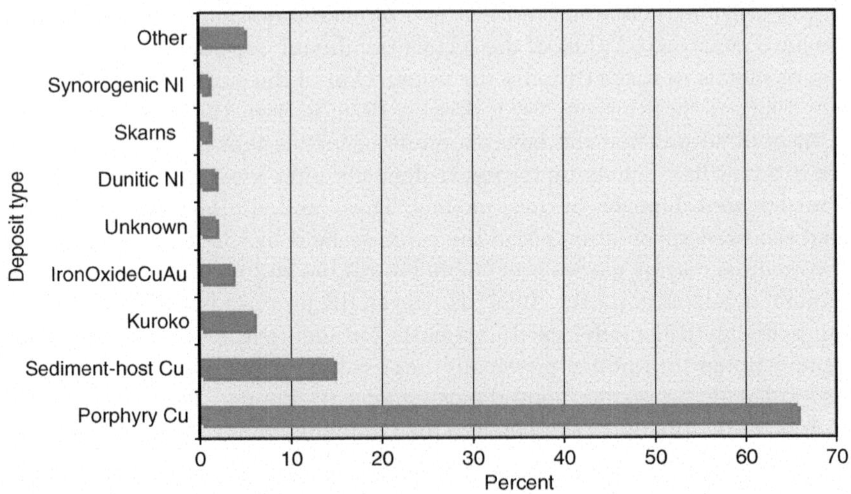

Figure 2.3 Percent of all known copper in each type of deposit.

silver in a 61-m cube, copper in a 620-m cube, zinc in a 500-m cube, and lead in a 340-m cube. The point is that, at least for these metals, the total amounts of metal presently known in deposits represent relatively small volumes, and they tend to be located in few places.

The known deposits, regardless of type, also can be used to examine whether the metals are predominately in lower grade deposits and whether many small deposits can be significant sources of metals. Comparison of the proportion of deposits with the proportion of metals grouped in increasing grade classes (average grades of whole deposits) provides a way to examine both frequency of grades and tendency of metals to concentrate at certain grades (figure 2.4). At least 62 percent of gold, silver, zinc, and lead is in deposits having average grades above the respective median grades, and 75 percent of copper is in deposits with average grades at or above a grade of 0.5 percent copper. That is, very low grade deposits contain less total metal than do moderate grade deposits. Comparison of the proportion of deposits with the proportion of metals grouped in increasing tonnage classes (tonnage of whole deposits) provides a way to examine both frequency of tonnages and tendency of metals to concentrate at certain tonnages (figure 2.5). Tonnage of mineralized rock is a better predictor of contained metal, with more than 96 percent of each metal's total residing in deposits having greater than median size and between 47 and 86 percent of metal contained in the largest 10 percent of deposits (figure 2.6).

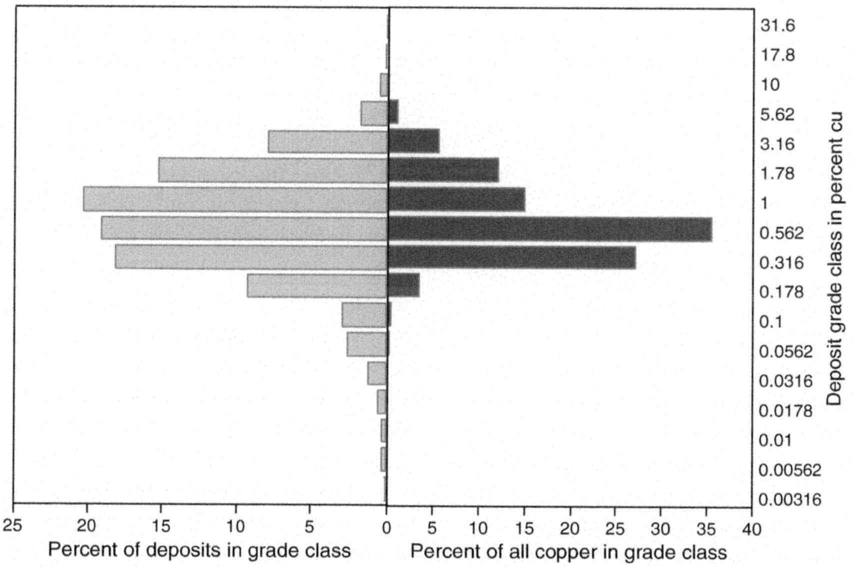

Figure 2.4 Percentage of all copper metal (right x-axis) and percentage of deposits (left x-axis) by average grade class (y-axis) for 2,045 copper-bearing deposits containing 2,065,000,000 metric tons of copper.

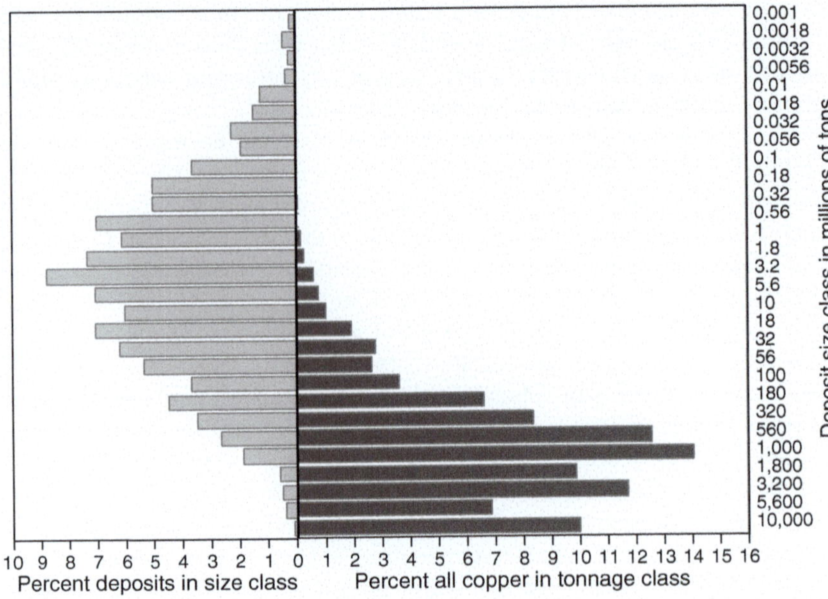

Figure 2.5 Percentage of all copper metal in deposits in size class (right x-axis) and percentage of deposits in tonnage class (left x-axis) by size of deposit class (y-axis) for 2,045 copper-bearing mineral deposits containing 2,065,000,000 metric tons of copper.

World-class deposits, defined as the upper 10 percent of deposits in terms of contained metal, account for more than 83 percent of all gold, 78 percent of silver, 83 percent of copper, 71 percent of zinc, and 77 percent of lead. Each of these world-class deposits contain at least 120 metric tons (3.2 million ounces) of gold, 2,400 metric tons (77 million ounces) of silver, 1.7 million metric tons copper, 1.6 million metric tons zinc, and 0.84 million metric tons lead. Mineral deposits occur rarely in the earth's crust, and large ones are especially uncommon (figure 2.6).

This analysis shows that only the unusually large deposits can significantly affect supply of these metals, and they are key to understanding exploration risks. The importance of effects of deposit size on estimating total resources and on exploration risk was demonstrated in a preliminary sensitivity analysis in which the expected amount of metal was estimated from a population of porphyry copper deposits (Singer and Kouda, 1999a). The greatest opportunity for reducing uncertainty and risk in exploration and resource assessment was shown to lie with lowering the uncertainty associated with tonnage estimates, followed in importance by uncertainty associated with grade estimates. The importance of the few largest deposits to mineral supply cannot be overemphasized—it is only the few largest that can affect supply.

ASSESSMENT EXAMPLE

In the 1990s the number of requests made to the USGS for mineral resource assessments and the complex nature of the requests made it difficult to respond in a timely manner. Both the boundaries of the areas to be assessed and deadlines for completed reports changed frequently. Preparing a detailed assessment of the nation in anticipation of future assessment requests was impractical. The USGS undertook a national mineral assessment of gold, copper, silver, lead, and zinc in order to provide a framework for the yet-to-be-requested more detailed assessments expected that have unknown boundaries and scales. Preparing the national assessment required organizing and consolidating large geological, geophysical, geochemical, and mineral deposit files that could then be used in subsequent assessments. This seminal quantitative assessment of the nation's resources used the three-part form and demonstrated that the United States contains about as much of the five metals in undiscovered conventional-type deposits as has already been found (for overview, see Schulz and Briskey, 2007).

Where might these large undiscovered deposits exist? In a USGS assessment of gold, silver, copper, zinc, and lead resources of the United States (Schruben, 2002), the majority of undiscovered mineral deposits were estimated to exist under cover. Because a large portion of undiscovered resources are covered by younger sediments or rocks, it is necessary to explicitly address the issue of predicting mineral resources remotely through cover in order to reduce uncertainties of possible locations and amounts.

Demand

The physical existence of minerals and the ability and willingness to produce minerals only partially determine whether minerals will be abundant or scarce. The level at which minerals are consumed is also important. Rapid economic development in Asia is increasing the consumption of mineral commodities and raises the question of whether mineral production will be adequate to meet the needs of society over the next twenty or more years. The answer to this question will in part depend on how fast developing countries increase their mineral consumption and whether developed countries are able to reduce the amount of minerals they consume, that is, in changes in emerging patterns of mineral consumption.

In theory, consumption of minerals can be measured at different stages of use, including as raw minerals by mineral processing industries, as

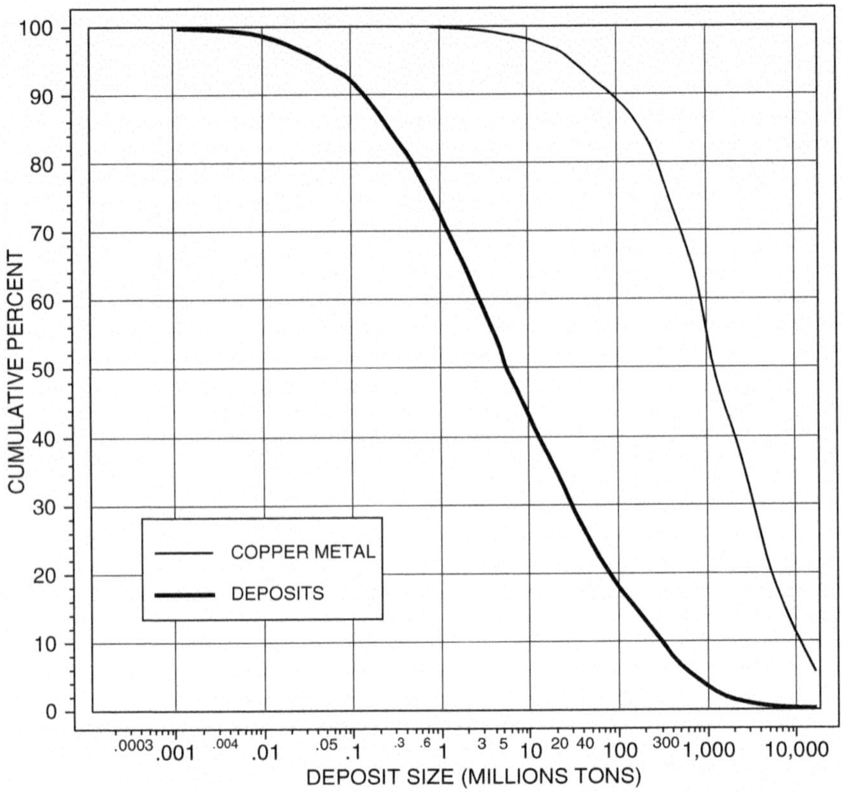

Figure 2.6 Cumulative percentage of 2,045 known copper-bearing deposits and percentage of their total amount of 2,065,000,000 metric tons of copper versus tonnage of mineralized rock. For example, deposits of 100 million tons or more represent about 18 percent of the deposits but 89 percent of the total copper.

processed minerals by industries, and as minerals in final goods used by consumers. Consumption is most commonly measured at the industrial stage because data on production and trade in primary metal products are easily obtained. Measuring mineral consumption in final goods is difficult because it requires disaggregating the finished goods into component materials, which can require a large investment of time and strong assumptions about the compositions of classes of goods.

Effects of a country's stage of development on mineral consumption has been examined by recent studies of changes in per capita apparent consumption of aluminum, cement, copper, and salt for Japan, the Republic of Korea, and the United States at 5-year intervals from 1965 to 1995 (DeYoung and Menzie, 1999) and of per capita apparent consumption data of the same four mineral products for the twenty most populous countries for every fifth year from 1970 to 1995 (Menzie, DeYoung, and Steblez, 2001). The study of Japan, Korea, and the United States (DeYoung and Menzie, 1999) found that, in the United States, per capita consumption of cement, copper, and salt did not significantly change over the period of study. In Korea, per capita consumption of aluminum, cement, and copper increased dramatically, while the consumption of salt grew slowly. In Japan, per capita consumption all four commodities grew at slow rates. Figure 2.7 illustrates changing per capita consumption of copper and cement in the three countries. By the end of the period, Japan and the United States had achieved a similar level of development as measured by per capita GDP (gross domestic product), and both countries consumed similar per capita levels of the metals, copper (10 kg), and aluminum (30 kg), whereas their levels of per capita consumption of cement and salt, a construction material and industrial mineral, respectively, differed by a factor of 2. The order in which growth began in Korea may be significant. Cement consumption grew earliest, followed by copper and aluminum. This could reflect development of basic infrastructure, followed by development of workplaces and light manufacturing, followed by manufacture of heavier structural goods.

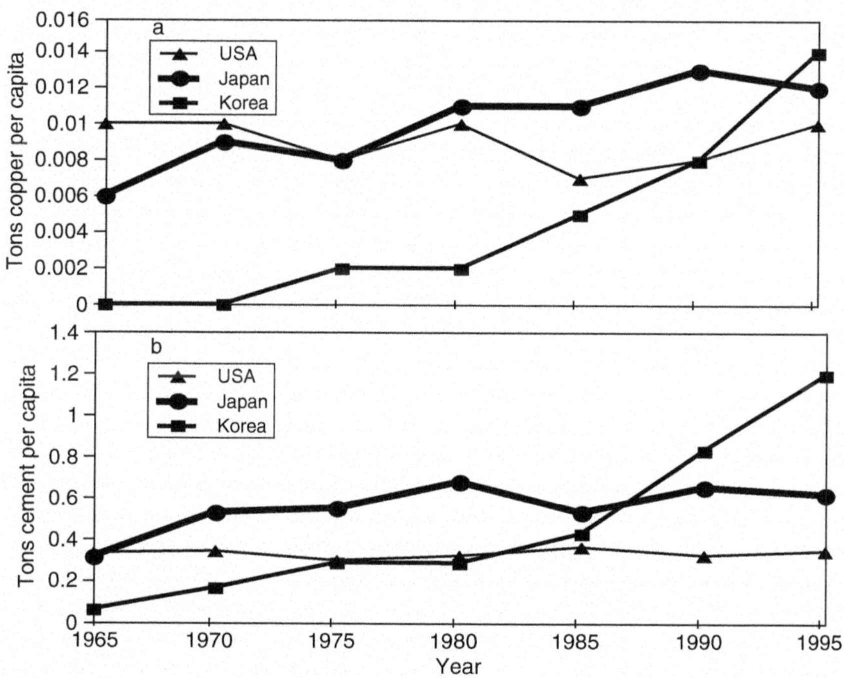

Figure 2.7 Copper (a) and cement (b) consumption per capita in Japan, Republic of Korea, and the United States. From DeYoung and Menzie (1999).

The second study (Menzie, DeYoung, and Steblez, 2001) included additional developed, or high-income, countries (Germany, United Kingdom, and France), which, like the United States and Japan, did not show significant changes in per capita consumption of cement and copper over relatively long periods but did show increasing per capita consumption of aluminum. Menzie, DeYoung, and Steblez (2001) also included countries that, like Korea, were undergoing significant development (increases in income) and significant increases in per capita consumption of aluminum, cement, and copper. China, Thailand, and Turkey showed significant increases in the per capita consumption of all three commodities, although Thailand's consumption of copper declined slightly in 2000 and Thai consumption of cement declined significantly in response to the Asian economic crisis of the late 1990s. Several other developing countries showed significant increases in one or two commodities: Mexico (aluminum and cement), Egypt (cement), and Iran (copper). Finally, two additional groups of countries showed other patterns of consumption. India and Indonesia also showed increases in per capita consumption of several mineral commodities, but because their initial level of consumption was low, growth is only beginning to appreciably increase their total consumption of minerals. Another group of countries, including Bangladesh and Nigeria, showed no significant consumption of mineral commodities.

Together, these two studies showed a consistent pattern of per capita consumption of minerals with increasing per capita income. At very low income levels, per capita consumption of minerals is very low. During the initial states of income growth, per capita consumption increases slowly. After income has reached a threshold level, per capita consumption increases very rapidly. Finally, when countries reach high levels of income, the per capita consumption is essentially constant. The pattern of use of these commodities is consistent with the phenomenon of dematerialization, or declining consumption per unit of economic activity (Wernick et al., 1996). The pattern, however, does not imply decreasing per capita consumption. Rather, results of the two studies suggest that levels of per capita consumption of metals in developed countries may be similar, while the per capita consumption of construction materials and industrial minerals may differ from country to country. Taken together, the pattern of increasing per capita consumption of minerals with increasing incomes and the similar absolute levels of use of metals form a basis for predicting future consumption of metals.

The pattern of increasing per capita consumption with increasing income described above defines a growth curve, which may be modeled by a logistic function (Menzie, Singer, and DeYoung, 2005):

$$C = (K/(1 + \mathrm{e}^{-r\log(i)}))\mathrm{P}, \tag{2.1}$$

where C is consumption of a commodity, K is a constant representing the saturation level of the per capita consumption of the commodity in an economy, r is a constant, i is per capita income or per capita GDP, and P is population. Thus, consumption of raw materials in the future may be calculated by adjusting per capita GDP for real growth of GDP using available estimates of population and income.

Because the consumption of copper in high-income countries became stable at about the same level, copper was selected to develop a model of consumption as a function of income and population. The equation was used to estimate future levels of copper consumption in the twenty most populous countries in 2020. The model forecasts copper consumption of the twenty countries will be 19 Mt (million metric tons) in 2020 (Menzie, Singer, and DeYoung, 2005). If the twenty countries consume the same proportion of world copper in 2020 as they did in 2000, world copper consumption in 2020 will be 24 Mt. This is 1.8 times larger than 2000 consumption and represents an average rate of growth of world copper consumption of 3.1 percent, which is about 10 percent faster than the annual growth rate (2.8 percent) of per capita consumption of copper between 1980 and 2000.

Developing countries have been increasing their proportion of world copper consumption. In 1980, France, Germany, Japan, the United Kingdom, and the United States accounted for 68 percent of world copper consumption. By 2000, these five countries accounted for only 51 percent of copper consumption. The model (equation 2.1) predicts that these five countries will account for only 30 percent of the world's consumption of copper by 2020. One way to comprehend the magnitude of changes implied by the model is to compare which countries consume more than 1 Mt of copper. In 1980, the Soviet Union consumed 1 Mt, Japan 1.2 Mt, and the United States 2.2 Mt. In 2000, Germany and Japan each consumed 1.3 Mt, China 2 Mt, and the United States 3 Mt. The model estimates that, in 2020, Brazil will consume 1.2 Mt, Japan 1.4 Mt, India 1.6 Mt, the United States 3.5 Mt, and China 5.6 Mt. New modeling indicates that China will be the world's leading consumer of industrial copper by 2015 and the twenty most populous countries will consume 36 Mt of copper in 2025. Such estimates may appear remarkable until one recalls the growth that Korea achieved in twenty years.

The above projected growth in consumption has important implications about the need to develop new reserves of copper and other minerals. The USGS reports world copper reserves in 2001 of 340 Mt (Edelstein, 2002). In 1995, world copper consumption was 10.5 Mt. Copper reserves are equal to thirty-four years of supply at the 1995 level of consumption. The logistic model (equation 2.1) predicts world consumption of more than 24 Mt per year by 2020. If consumption reaches this level and reserves are kept proportional to consumption, at about thirty years of consumption, and allowance is made for intervening drawdown of reserves, more than 700 Mt of copper must be added to world reserves, assuming that 10 percent of consumption will come from secondary sources. The likelihood of developing the reserves indicated by the above calculations depends on the rate of discovery of large deposits. Porphyry copper and sedimentary copper deposits currently

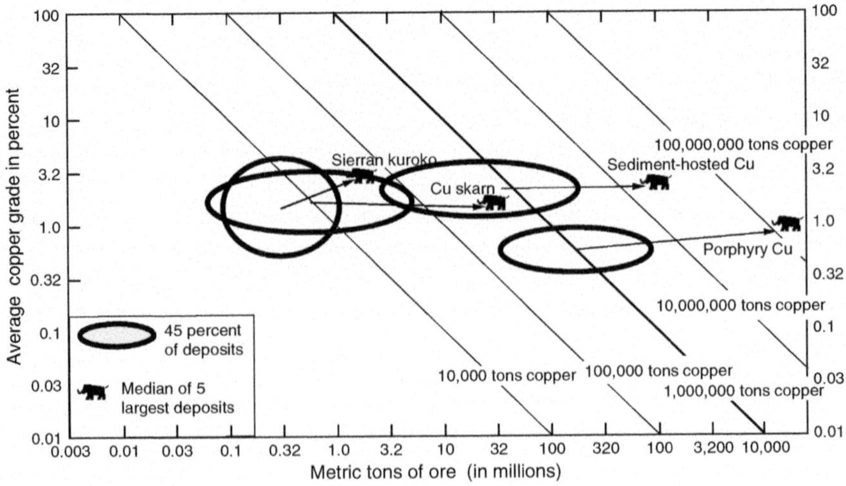

Figure 2.8 Copper grades, tonnages, and contained metal by type of deposit.

contain the bulk of the world's copper reserves. The ability to discover these additional reserves depends on new discoveries of large porphyry copper and/or sedimentary copper deposits and/or discoveries of significant reserves in a major new type of copper deposit. Typically, porphyry copper deposits contain more copper than sedimentary copper deposits (figure 2.8). The five largest porphyry copper deposits contain a median of about 100 Mt of copper. In order to maintain the predicted growth in consumption and retain a thirty-year reserve, the discovery of 1.1 Gt of copper in new deposits will be required by 2020. To put this immense amount into perspective, it will require the discovery and development of eleven additional deposits, each of which is as large as the median of the five largest porphyry copper deposits known from a century of exploration!

Implications

The rapid economic development in Asia and increase in consumption of mineral commodities that is accompanying this development are reviving concerns of about resource scarcity. Will mineral production be adequate to meet the needs of rapid economic development of large Asian countries? At the same time, there are growing concerns that global environmental systems are already being affected by mineral production and use. Can global environmental systems absorb the wastes from increased mineral production and use, or can we reduce the levels of such wastes sufficiently to avoid environmental problems?

Based on estimates of known and yet-to-be-discovered mineral resources, mineral supply per se does not appear to be a limit on growth. In addition, mineral deposits that occur at, or near, the surface are still being found in some regions. Over the next twenty or more years, however, an increasing proportion of resources that remain to be discovered are likely to be in deposits concealed beneath significant quantities of covering rock or sediments. Such deposits will be more difficult to discover and will likely be more costly to mine.

Very large deposits are needed to sustain supply and are more likely than small deposits to have high net present values (Singer, 2005). Economic trends of lower prices over time require larger sized deposits in order to take advantage of economies of scale. Many major exploration companies today require a minimum size of what is called "world-class" deposits. Even in parts of the world where exploration of exposed rocks is only partially complete, the belief that the larger deposits tend to be found early in the exploration process suggests that the new frontier of exploration lies under cover where there is an increased chance of discovering world-class deposits. The possibility of large returns and a substantial risk of failure argue for careful planning and execution of exploration under cover. Because of high costs of exploring under cover, it is critical to plan exploration strategies that efficiently use models, experienced geoscientists, and data.

For both governmental and industrial decision-makers, utilizing information necessary to develop policies to reduce disruptive fluctuations in the supply of minerals requires understanding how these critical factors can affect supply. One of the effects of globalization is that both suppliers and consumers have an interest in possible supplies of mineral materials. Understanding that significant mineral supply typically comes from rare large deposits of a few types that do not occur everywhere is key to proper planning. Continuing growth of demand in large developing countries and increasing environmental concerns suggest continued price pressures of mineral raw materials in the next ten years or so. These supplies will largely come from existing or soon-to-open mining operations. Replenishing the materials mined and supplying additional minerals require further development of discovered mineral deposits and the discovery of new, yet-to-be-identified deposits. Critical to exploration decisions must be the awareness of the need for very large deposits, the large expense implied by remote or covered deposits, and the time lag in finding and bringing such deposits to the point of production. It is the information from quantitative mineral resource assessments of these yet-to-be-discovered deposits that benefits decision-makers by reducing uncertainty and allowing better informed decisions. Information from quantitative assessments allows governmental and company planners to consider whether and when to invest in exploration to meet future mineral needs.

For mineral resource assessors, information required to make future assessments necessitates dealing with quantitative data, with economics, and, in many situations, with information necessary to assess mineral resources that are covered or in remote places. Assessing resources under cover will prove particularly difficult because of the common lack of prospects and of mapped geology. Without maps showing boundaries of geologic settings, there is no basis to define boundaries of where different kinds of mineral deposits might occur. Although only deposit types that have large tonnages will be the focus of most assessments because of their possible effect on supply, there are situations where smaller types of deposits might be of interest because of their association with larger deposit types. In addition, some deposit types that are typically small in size, such as some polymetallic veins that might contain such rare elements as indium, could be targets of interest. Most forms of mineral resource assessment do not provide the information necessary to address these issues (Singer and Mosier, 1981a; Harris, 1984; Shulman et al., 1992). Finally, the three-part form of mineral resource assessments presented in this book is designed to show how to integrate the information so that assessors will be able to better communicate the results to decision-makers.

3

Descriptive Models

Perspective

Mineral deposit models play a central role in an information system that will help the policy makers to make their decisions. Ideally, the different kinds of deposit models would provide the necessary and sufficient information to discriminate (1) possible mineralized environments from barren environments, (2) types of known deposits from each other, and (3) mineral deposits from mineral occurrences. Probably the most important part of creating mineral deposit models is the planning stage in which consideration of the purpose and possible uses of the models should determine the character of the models. The way to describe a model is first by thinking about what it is for, about its function, not the list of items that make up its structure (Churchman, 1968).

Although there are many fine compendiums of mineral deposit models (Australian Geological Survey Organisation, 1998; Eckstrand, Sinclair, and Thorpe, 1995; Kirkham et al., 1993; Lefebure and Hoy, 1996; Lefebure and Ray, 1995; Roberts and Sheahan, 1988; Rongfu, 1995; Sheahan and Cherry, 1993), the focus in this book is on deposit models applied to quantitative resource assessment. The focus of this chapter is the descriptive aspects of the deposits because the goal is to provide a basis for interpreting geologic observations rather than to provide interpretations in search of examples (Cox, Barton, and Singer, 1986). Thus, the discussion herein is limited to mineral deposit models specifically designed for quantitative assessments such as those in Cox and Singer (1986), Bliss (1992a), Orris and Bliss (1991, 1992), and Rogers et al. (1995). Mineral deposits modeled for three-part assessments are defined as mineral occurrences of sufficient size and grade that they might, under favorable circumstances, be economic. Although history suggests that we can expect discoveries of as-yet-unrecognized deposit types, the three-part assessments discussed here do not include resources from these deposits simply because they cannot be modeled.

Most published quantitative mineral resource assessments that have used models have relied upon descriptive and grade-and-tonnage models (chapter 6), which are also the foundations of other kinds of models such as deposit-density models (chapter 4) and economic cost models (chapter 5). One of the purposes of a mineral deposit model is to communicate information that helps us find and evaluate mineral deposits. In general, a mineral deposit model is the systematically arranged information describing the essential attributes (properties) of a class of mineral deposits (Barton, 1993).

ASSESSMENT EXAMPLE

An assessment of the mineral resources of Colombia (Hodges et al., 1984) was the first international quantitative resource assessment by the U.S. Geological Survey (USGS). It was performed jointly with the Colombian agency INGEOMINAS in order to encourage development of mineral resources. Documentation and development of mineral deposit models was required in order to communicate with colleagues who spoke another language and had limited access to the latest literature on economic geology. Thus, the first descriptive mineral deposit models designed specifically for assessments were born. Although there had been earlier grade-and-tonnage models, the models created with this assessment were the first to accompany descriptive models and include the information in graphical form. Following publication of these deposit models in 1983, models specifically on Canadian deposits were published (Eckstrand, 1984), and the compilation of worldwide models was published by the USGS (Cox and Singer, 1986).

Because every mineral deposit is different from every other in some way, models have to represent more than single deposits. Deposits sharing a relatively wide variety and large number of attributes come to be characterized as a "type," and a model representing that type can be synthesized. Deposit models are constructed with information in and around know deposits, and as a consequence, the models contain information that can be used to discriminate one deposit type from another. For example, low values of an attribute (X) (figure 3.1) can be used to clearly discriminate deposit type d_1 from deposit type d_2 because low values of X have essentially zero probability

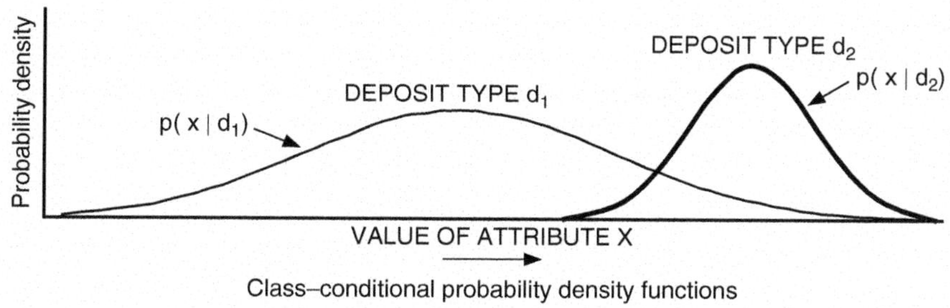

Figure 3.1 Probability of observing a value of attribute X, given the sample came from deposit type d_1 or type d_2.

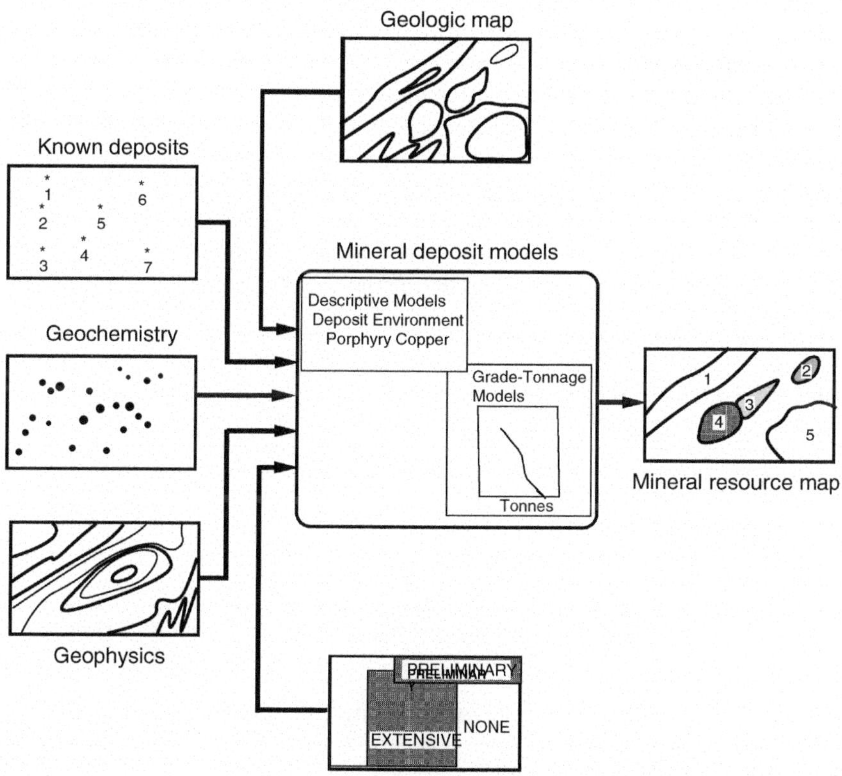

Figure 3.2 Diagram showing the central role of mineral deposit models in integrating information to delineate tracts.

of occurring given deposit type d_2. But attribute X does not necessarily discriminate a barren area from either deposit type d_1 or d_2. In quantitative assessments, deposit models are used to classify mineralized and to classify types of known deposits in the first, tract-delineation part of the assessment (figure 3.2), and mineral deposits are distinguished from mineral occurrences in the third, number-of-deposit estimation part of the assessment. The second part of assessments, grade-and-tonnage models of deposits, is combined with the estimated number of undiscovered deposits to provide the foundation for economic analysis.

Descriptive models used in three-part assessments focus on observations and use theories of origin only to guide what to observe. Ideally, the observations are available at the scale of the assessments. Descriptive models, such as those in Cox and Singer (1986), have two parts (tables 3.1, 3.2). The first describes the geologic environments in which the deposits are found; the second gives identifying characteristics of deposits. The second part of the descriptive model, the deposit description, provides the identifying characteristics of the deposits themselves, particularly emphasizing aspects by which the deposits might be recognized and used to discriminate one type from another, such as mineralogy, alteration, and geochemical and geophysical anomalies.

Thus, the first part of a descriptive model describes the general setting of the deposit type and plays a primary role in the delineation of tracts of land geologically permissive for the occurrence of undiscovered deposits. This information may be found on typical regional-scale geologic maps. The second part helps classify known deposits and occurrences into types, which aids the delineation process discussed in chapter 7. In some cases, the typing of known deposits and occurrences helps to identify geologic environments

Table 3.1. Example of a descriptive model of porphyry copper-gold deposits by Cox (1986a).

Description Stockwork veinlets of chalcopyrite, bornite, and magnetite in porphyritic intrusions and coeval volcanic rocks. Ratio of Au (ppm) to Mo (percent) is greater than 30.

General References Sillitoe (1979), Cox and Singer (1992).

Geological Environment

Rock Types Tonalite to monzogranite; dacite, andesite flows and tuffs coeval with intrusive rocks. Also syenite, monzonite, and coeval high-K, low-Ti volcanic rocks (shoshonites).

Textures Intrusive rocks are porphyritic with fine- to medium-grained aplitic groundmass.

Age Range Cretaceous to Quaternary.

Depositional Environment In porphyry intruding coeval volcanic rocks. Both involved and in large-scale breccia. Porphyry bodies may be dikes. Evidence for volcanic center; 1–2 km depth of emplacement.

Table 3.1. (*continued*)

Tectonic Setting(s) Island-arc volcanic setting, especially waning stage of volcanic cycle. Also continental margin rift-related volcanism.
Associated Deposit Types Porphyry Cu-Mo; gold placers.

Deposit Description

Mineralogy Chalcopyrite ± bornite; traces of native gold, electrum, sylvanite, and hessite. Quartz + K-feldspar + biotite + magnetite + chlorite + actinolite + anhydrite. Pyrite + sericite + clay minerals + calcite may occur in late-stage veinlets.
Texture/Structure Veinlets and disseminations.
Alteration Quartz ± magnetite ± biotite (chlorite) ± K-feldspar ± actinolite, ± anhydrite in interior of system. Outer propylitic zone. Late quartz + pyrite + white mica ± clay may overprint early feldspar-stable alteration.
Ore Controls Veinlets and fractures of quartz, sulfides, K-feldspar magnetite, biotite, or chlorite are closely spaced. Ore zone has a bell shape centered on the volcanic-intrusive center. Highest grade ore is commonly at the level at which the stock divides into branches.
Weathering Surface iron staining may be weak or absent if pyrite content is low in protore. Copper silicates and carbonates. Residual soils contain anomalous amounts of rutile.
Geochemical Signature Central Cu, Au, Ag; peripheral Mo. Peripheral Pb, Zn, Mn anomalies may be present if late sericite pyrite alteration is strong. Au (ppm):Mo (percent) > 30 in ore zone. Au enriched in residual soil over ore body. System may have magnetic high over intrusion surrounded by magnetic low over pyrite halo.

Examples

Dos Pobres, Arizona (Langton and Williams, 1982)
Copper Mountain, British Columbia, Canada (Fahrni et al., 1976)
Tanama, Puerto Rico (Cox, 1985)

Table 3.2. Example of a descriptive model of porphyry copper deposits by Cox (1986b).

Description This generalized model includes various subtypes all of which contain chalcopyrite in stockwork veinlets in hydrothermally altered porphyry and adjacent country rock.
General Reference Titley (1982).

Geological Environment

Rock Types Tonalite to monzogranite or syenitic porphyry intruding granitic, volcanic, calcareous sedimentary, and other rocks.
Textures Porphyry has closely spaced phenocrysts and microaplitic quartz-feldspar groundmass.

(*continued*)

Table 3.2. *(continued)*

Age Range Mainly Mesozoic and Cenozoic, but may be any age.
Depositional environment High-level intrusive rocks contemporaneous with abundant dikes, breccia pipes, faults. Also cupolas of batholiths.
Tectonic setting(s) Rift zones contemporaneous with Andean or island-arc volcanism along convergent plate boundaries. Uplift and erosion to expose subvolcanic rocks.
Associated deposit Types Base-metal skarn, epithermal veins, polymetallic replacement, volcanic hosted massive replacement. See also: Porphyry Cu-skarn related, porphyry Cu-Mo, and porphyry Cu-Au.

Deposit Description
Mineralogy Chalcopyrite + pyrite ± molybdenite; chalcopyrite + magnetite ± bornite ± Cu; assemblages may be superposed. Quartz + K-feldspar + biotite + anhydrite; quartz + sericite + clay minerals. Late veins of enargite, tetrahedrite, galena, sphalerite, and barite in some deposits.
Texture/Structure Stockwork veinlets and disseminated sulfide grains.
Alteration From bottom, innermost zones outward: sodic-calcic, potassic, phyllic, and argillic to propylitic. High-alumina alteration in upper part of some deposits. Propylitic or phyllic alteration may overprint early potassic assemblage.
Ore Controls Stockwork veins in porphyry, along porphyry contact, and in favorable country rocks such as carbonate rocks, mafic igneous rocks, and older granitic plutons.
Weathering Green and blue Cu carbonates and silicates in weathered outcrops, or where leaching is intense, barren outcrops remain after Cu is leached, transported downward, and deposited as secondary sulfides at water table or paleowater table. Fractures in leached outcrops are coated with hematitic limonite having bright red streak. Deposits of secondary sulfides contain chalcocite and other Cu_2S minerals replacing pyrite and chalcopyrite. Residual soils overlying deposits may contain anomalous amounts of rutile.
Geochemical Signature Cu + Mo + Au + Ag + W + B + Sr center; Pb, Zn, Au, As, Sb, Se, Te, Mn, Co, Ba, and Rb outer. Locally Bi and Sn form most distal anomalies. High S in all zones. Some deposits have weak U anomalies.

Examples
Bingham, Utah (Lanier et al., 1978)
San Manuel, Arizona (Lowell and Guilbert, 1970)
El Salvador, Chile (Gustafson and Hunt, 1975)

not indicated on geologic maps. The arrangement of descriptive models illustrated here is designed to focus on host-rock lithology and tectonic setting, the features most easily obtained from a geologic map. Experimental forms of descriptive models emphasizing quantitative information are discussed later in this chapter.

General Geologic Setting

The section titled "Geological Environment" in tables 3.1 and 3.2 provides information about the geologic environment under several headings. The headings "Rock Types" and "Textures" describe the host rocks of deposits as well as the source rocks believed responsible for some deposits. In table 3.1, the rocks tonalite to monzogranite and dacite and andesite flows and tuffs coeval with intrusive rocks are listed for the gold-rich variety of porphyry copper deposits (Cox and Singer, 1992). Also listed are syenite, monzonite, and coeval high-potassium, low-titanium volcanic rocks. In the more general porphyry copper model (table 3.2), syenitic porphyry intruding granitic, volcanic, calcareous sedimentary, and other rocks has been added. Textures with these deposits are listed as porphyritic intrusive rocks with fine- to medium-grained aplitic groundmass.

Some of these rocks and textures, such as porphyry or breccia, are only identifiable on quite detailed maps due to the small aerial size of the bodies—this situation occurs in many deposit models. The contents of the descriptive model might be correct but are not useful at the scale of some mineral resource assessments. In general, scales of observations in the "Geological Environment" section of descriptive models have not been identified, and yet scales are important in the application of these models in assessments because differences in map scale frequently reflect differences in the information content of the maps.

"Age Range" refers to the age of the event responsible for the formation of the deposit. "Depositional Environment" refers to the geologic setting of the deposit. "Tectonic Setting" is concerned with major tectonic features or provinces. "Associated Deposit Types" are listed as deposit types whose presence might indicate suitable conditions for the formation of deposits of the type portrayed by the model. For the descriptive model for porphyry Cu-Au (table 3.1), the associated deposit types are listed as porphyry Cu-Mo and gold placers. In the more general porphyry copper model (table 3.2), the associated deposits are listed as base-metal skarn, epithermal veins, polymetallic replacement, and volcanic hosted massive replacement. The "Geological Environment" section of the descriptive model uses information from the geologic map, geophysical maps, and the known deposits and occurrences. This part of a descriptive model provides the general setting of the deposit type and plays the primary role in the delineation of tracts of land geologically permissive for the occurrence of undiscovered deposits (see chapter 7).

ASSESSMENT EXAMPLE

In 1994, at the request of the U.S. State Department, two USGS geologists went to Jakarta to consult with Indonesian government officials and to plan a possible resource assessment project. The Indonesian government was particularly interested in mineral resources in the eastern part of the country in order to possibly provide employment opportunities for people relocating from the densely populated western part of the country. The goal of the proposed project was to prepare a quantitative assessment of eastern Indonesia's mineral resources that would help the government of Indonesia develop its mineral resources to ensure sustainable development and in ways that are sensitive to environment. Before writing the proposal for the project, a short course was given and a field trip was taken to examine the geology of some parts of eastern Indonesia. Among the places visited was Ambon Island, which was believed to have low chances of containing metal-bearing economic minerals and that had no known prospects or occurrences. The geologists found clear indications of the kind on mineralization associated with porphyry copper-gold deposits (Menzie et al., 1997). Subsequent exploration by a Canadian mining company confirmed the possibility of a large copper deposit, and Indonesian government geochemical work suggested the possibility of associated epithermal gold deposits on Ambon Island. Unfortunately, due to subsequent violence, further exploration has not taken place on Ambon, and the joint assessment project was not funded.

According to the descriptive models of Cox (1986a, 1986b), porphyry copper deposits consist of stockwork, disseminated, and breccia-hosted copper mineralization together with K-silicate alteration that is generally restricted to porphyritic stocks and their immediate wall rocks. The deposits may have parts that contain skarn. Deposits that may be derived from, or affected by, supergene processes are included in the models. Using these descriptive models and an understanding of tectonic settings, broad volcanic arcs that formed at approximately the same time can be the fundamental unit for the delineation of permissive areas for porphyry copper deposits. Permissive tracts can be outlined along borders of magmatic arcs, taking into consideration the deposit ages and distributions of major structures. Boundaries of mapped rock units form the primary basis for drawing limits of permissive areas (see chapter 7). Permissive area boundaries are typically extended using interpolated geology and locally geophysical surveys, such as aeromagnetics, to identify where younger rocks or sediments conceal permissive rocks.

Deposit Description

The purpose of the "Deposit Description" section of these descriptive models is to assist in the classification of known mineral deposits and more poorly explored mineral occurrences into the appropriate deposit types. Properly classed deposits and occurrences are useful when plotted on a map in that they reinforce the spatial distributions of geologic settings of particular deposit types and, in some cases, identify geologic settings not recognized from the geologic map alone due to errors or to scale issues where the map scale is not sufficient to differentiate the permissive hosting unit (chapter 7).

"Mineralogy," "Texture/Structure," and "Alteration," which are frequently useful in classification, are included in the deposit description section of descriptive models for that reason. Although many deposit types have overlapping minerals in common, it is possible to correctly classify many deposits into types with mineralogy (see below). Information on weathering of deposits and their alteration patterns is applicable in many parts of the world. "Geochemical Signature" and, in some cases, geophysical anomalies (table 3.1; Hoover, Heran, and Hill, 1992), might be useful in planning for mineral assessments or mineral exploration (Cox, Barton, and Singer, 1986). However, unless the assessment is being done at a detailed map scale, such information may not be useful because it is not typically shown on regional maps. Many descriptive models also include a generalized map or cross section illustrating ore controls, zoning patterns, or other features of the model.

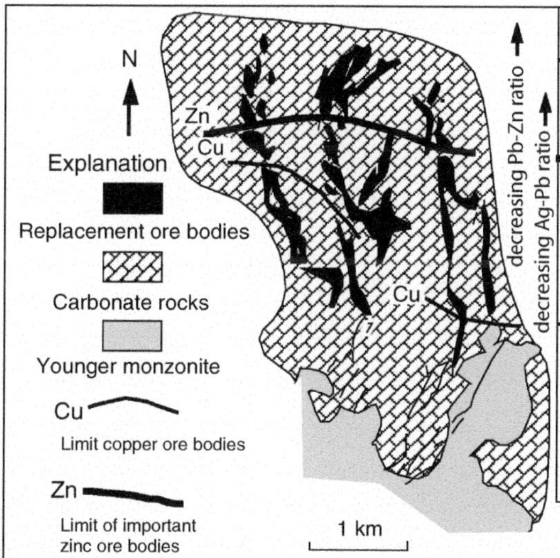

Metals	mineralogy: Ore	mineralogy: Gangue
Zn-Mn	sphalerite + rhodochrosite	fine-grained pyritic jasperoid
Pb-Ag	galena + sphalerite + argentite + Ag sulfosalts ± tetrahedrite	fine-grained jasperoid with barite crystals and quartz - lined vugs
Cu-Au	enargite - famatinite + tennantite - tetrahedrite + sphalerite + argentite + digenite	medium-grained jasperoid and coarse crystals of quartz and barite

Figure 3.3 Map illustrating a generalized model of metal and mineral zoning in polymetallic replacement deposits. After Morris (1986).

The generalized map or cross section is primarily used to classify known deposits and occurrences in and near an area being assessed (figure 3.3). In appendices to USGS Bulletin 1693 (Cox and Singer, 1986), the models are extensively indexed by frequency of occurrence of minerals and associated geochemical anomalies. The 3,900 individual deposits located in 110 countries referred to in the models in USGS Bulletin 1693 are also indexed, and their model classification is shown.

Digital Models

Consistency in quantitative assessments is dependent on the internal consistency required in the construction of the descriptive, grade-and-tonnage, and deposit-density models. Descriptive models discussed above have been developed on the basis of expert knowledge. An alternative, more time-consuming method of developing descriptive models is to gather data from well-explored deposits of each type to determine how commonly different attributes and combinations of attributes occur after expert classification of the deposits. Quantifying mineral deposit attributes is the necessary and sufficient next step in statistically classifying known deposits by type. Quantified deposit attributes also can provide a firm foundation to identify which observations on geologic and other maps should be effective in delineation of tracts and perhaps identifying sites for detailed exploration.

An example of these kinds of digital data is shown in figure 3.4, where the frequency of reports of the presence or absence a few of the minerals present in five types of deposits has been counted. Adularia and gold are more commonly reported in Comstock epithermal gold-silver deposits, whereas alunite and barite are more commonly reported in quartz-alunite epithermal gold-silver deposits. Biotite is ubiquitous in porphyry copper deposits but not particularly common in the other deposit types plotted. To determine if quantified mineral deposit models would be useful in classifying unknown deposits into types, data on the minerals reported present in fifty-five different types of deposits were compiled (Singer et al., 1997) and were used to statistically discriminate eight of the deposit types. Using fifty-eight minerals and six generalized rock types, 88 percent of the deposits not used in training the neural network were accurately classified into the correct eight types (Singer and Kouda, 1997a). Just knowing whether there are marine mafic volcanic rocks or marine felsic to intermediate volcanic rocks near the deposits increased the correctly classified deposits in the validation set from 88 percent to 98 percent. Clearly, digital mineralogy and simple rock types can be useful in classifying well-studied deposits. Using the same kind of information, the rate of correct classification was only about 53 percent for mineral prospects (occurrences only partially explored) (Singer and Kouda, 1997b). Typically, there is sparse information about mineralogy and even associated rock types of prospects. The success of expert economic geologists

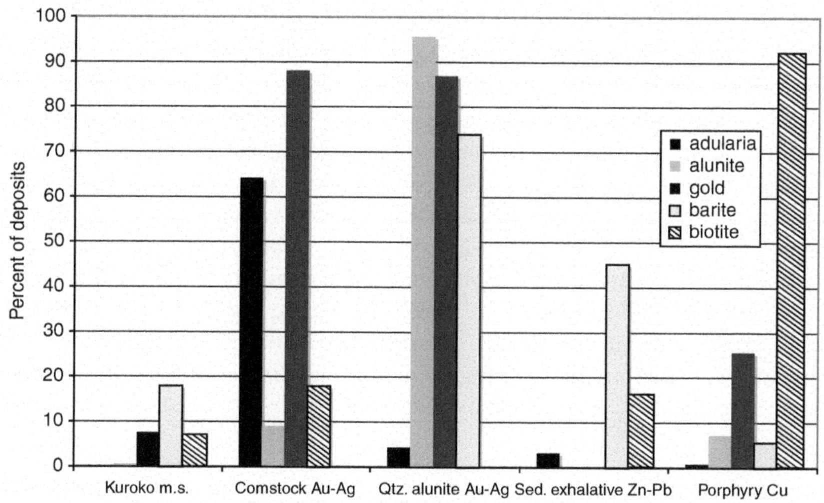

Figure 3.4 Percentages of deposits of selected types with indicated minerals reported.

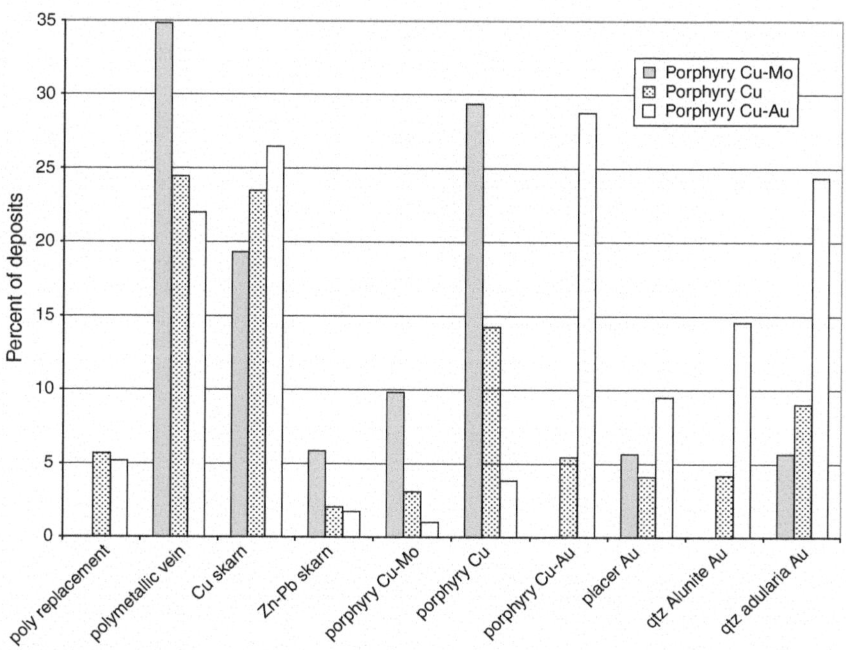

Figure 3.5 Percentages of selected deposit types observed within 10 km of three subtypes of porphyry copper deposits.

in classifying prospects depends on such information, and additionally they use information about spatially associated deposits and prospects that are classified.

Digital information on the spatial association of deposits is only available for a few deposit types. The pattern that emerges when one looks at the proportion of deposit types spatially associated with the subtypes of porphyry copper deposits strongly reinforces the suggested value of such information (figure 3.5). Epithermal quartz-alunite gold-silver, epithermal Comstock gold-silver, and polymetallic replacement zinc-lead are more common near the relatively shallow gold-rich porphyry copper deposits, but zinc-lead skarn deposits are more common near molybdenum-rich porphyry copper deposits. In addition, some of the deposit types that are relatively common

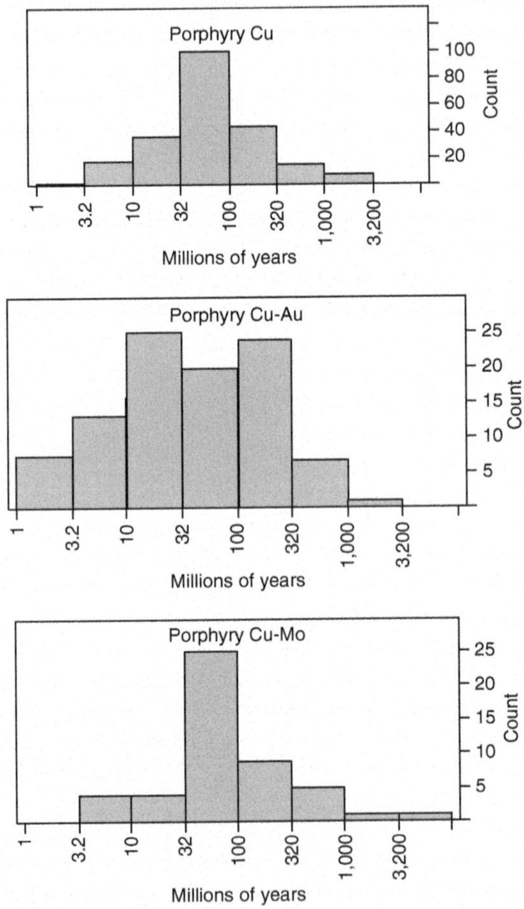

Figure 3.6 Ages of porphyry copper deposits by subtype.

within 10 km of porphyry copper deposits, such as polymetallic vein and quartz adularia epithermal vein gold-silver types, were not mentioned in the descriptive models of Cox and Singer (1986).

Digital models offer the advantages over expert-created descriptive models in that the information is documented and reproducible and can be used in classification and prediction (Singer, 2006). The advantages of digital data are evident if the frequency distribution of ages observed in subtypes of porphyry copper deposits (figure 3.6) are compared to the age of gold-rich porphyry copper deposits indicated in the descriptive model shown in table 3.1—clearly, these deposits as shown by these data have a wider age range than estimated by Cox (1986a).

The kind of digital models presented here about attributes associated with known deposits is necessary but not sufficient to discriminate barren from mineralized environments; quantifying the attributes of barren environments also is necessary for this task. In other words, part of the nature of descriptive models, whether made by experts or by using digital data, is that these models do not include information about barren areas.

4

Deposit-Density Models

Perspective

A key function of many forms of quantitative mineral resource assessments is estimation of the number of undiscovered deposits. In any given region, there is some fixed but, in most cases, unknown number of undiscovered deposits of a given type—the number could be zero or a larger integer. Many quantitative resource assessments that are based on a common three-part form of assessment (Singer, 1993a) have used expert judgment to estimate the number of deposits. Estimates of this unknown number are presented in a probabilistic form to reflect the uncertainty associated with the estimate.

Ideally, estimates of number of deposits should rely on analogies with similar well-explored geologic settings, just as grades and tonnages of well-explored deposits serve as analogs of the qualities and sizes of undiscovered deposits. Estimates of the number of undiscovered deposits can be derived from counts of known deposits per unit area in explored control regions. Number of deposits per unit area of the control regions can be used in histograms to show variation of densities by deposit type. Some research has been conducted on densities of several deposit types so that these ratios can be more widely used as a guide for number-of-deposit estimates (Bliss, Orris, and Menzie, 1987; Bliss, Menzie, Orris, and Page, 1987; Bliss and Menzie, 1993; Bliss, 1992b; Root, Menzie, and Scott, 1992). Most of these studies provide point (i.e., single) estimates of the number of deposits per unit area. Singer et al. (2001) summarize the ideas behind these mineral deposit-density models and provide individual estimates for twenty-seven combinations of deposit types and control locations. Many of the specially selected areas they describe provide standards to identify what should be considered high estimates of number of undiscovered deposits in most situations. Thus, many published mineral-deposit densities provide guides that suggest upper limits to estimates but are not necessarily useful in providing estimation guides for more likely situations. Four studies that attempt to address issues related to variability of deposit densities within a deposit type along with questions about effects of map or assessment scale are for low-sulfide quartz-gold veins (Bliss and Menzie, 1993), podiform chromite deposits (Singer, 1994a), porphyry copper type-deposits (Singer, Berger, Menzie, and Berger, 2005), and volcanogenic massive sulfide deposits (Mosier, Singer, and Berger, 2007). Powerful estimators of the density of deposits, number of deposits, and amount of mineralized rock based on analysis of ten different types of mineral deposits (Climax Mo, Cuban Mn, Cyprus massive sulfide, Franciscan Mn, kuroko massive sulfide, low-sulfide quartz-Au vein, placer Au, podiform Cr, porphyry Cu, and W vein) from 108 permissive control tracts around the world have been used to generalize across deposit types (Singer, 2008).

Deposit Densities

There are no fixed methods for making estimates of the number of undiscovered deposits. On the basis of experience and logic, however, a number of techniques can be used directly or as guidelines to make these estimates. Each method represents some form of analogy. Most robust of these methods is a form of mineral deposit model wherein the numbers of deposits per unit area from well-explored control regions are counted for a deposit type and the resulting frequency distribution is used either directly for an estimate or indirectly as a guideline in some other method. Figure 4.1 presents a hypothetical situation where, for an arbitrary deposit type, the number of discovered deposits is counted in each of twelve well-explored control tracts that are permissive for the deposit type, and the areas of each tract are recorded. Of course, the distribution of the deposit sizes is consistent with the associated grade-and-tonnage model.

Ratios of number of deposits per area are histograms showing how commonly different deposit densities are present (figure 4.2). It is relatively easy to determine if an area is well explored for some deposit types, such as porphyry copper or placer gold, if the deposits are exposed and the area is not heavily vegetated. Because of the difficulty of recognition of some types of deposits, such as sediment-hosted copper, determining the extent and

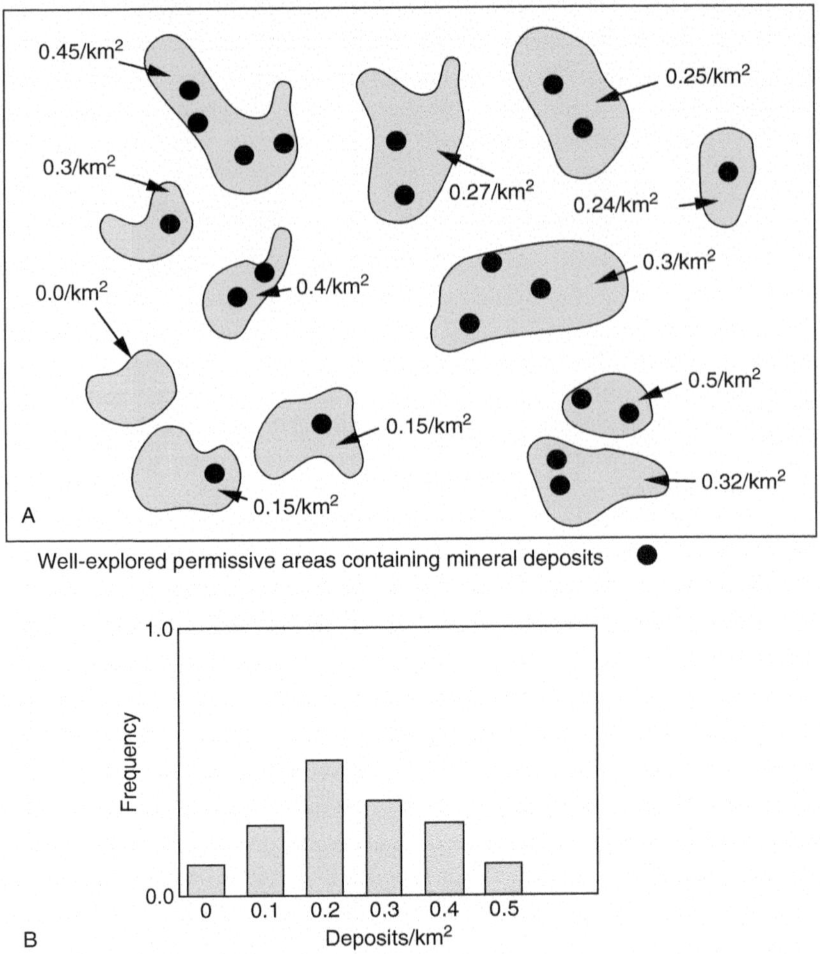

Figure 4.1 Example of known deposits in well-explored permissive areas (A), and histogram of derived deposit densities (B).

efficiency of exploration in an area is more difficult for these deposit types. It is not necessary that the control areas be explored completely, but it is necessary that the number of deposits found and the proportion of the area explored be estimated. An example of such an adjustment for incomplete exploration is presented below for kuroko massive sulfide deposits in the Hokuroku district of Japan. In some situations it is possible to consider mineral deposit density as the probability that a deposit of a given type occurs within some standard measure of area such as kilometer. We do not use that approach here because it requires a strong assumption that there can be one and only one deposit within the area.

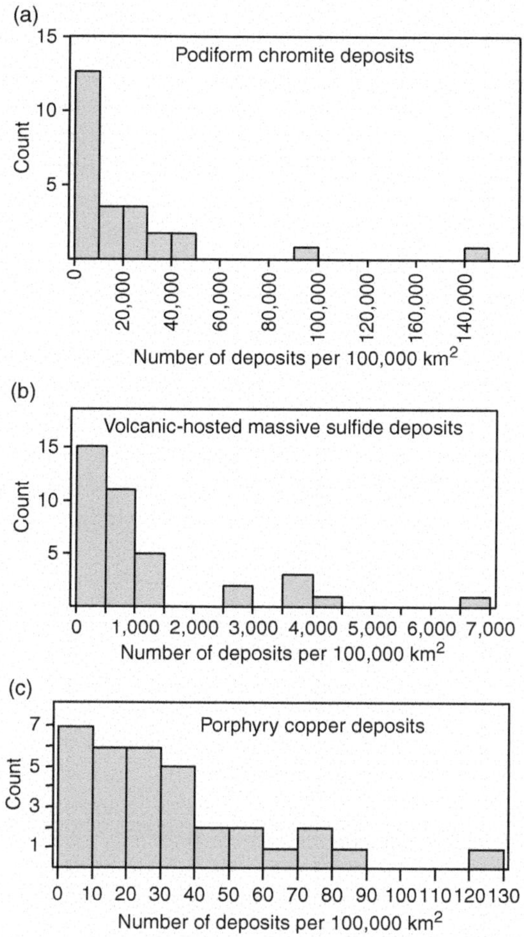

Figure 4.2 Histograms of the number of podiform chromite (a), volcanic-hosted massive sulfide (b), and porphyry copper (c) deposits per 100,000 km^2 per belt. Data are from Singer (2008).

Density Estimation

Deposit-density models have been designed to be used within the three-part form of assessment, which affects how the models should be constructed. In this kind of assessment, grade-and-tonnage models have the form of frequency distributions of tonnages and average grades of well-explored deposits of each type—they serve as models for grades and tonnages of undiscovered deposits of the same type occurring in geologically similar settings. In three-part assessments, estimates of number of undiscovered deposits explicitly represent the probability (or degree of belief) that some fixed but unknown number of undiscovered deposits that are consistent with the grade-and-tonnage model exists in the delineated tracts. Biases can be introduced into the resource estimates either by a flawed grade-and-tonnage model or by lack of consistency of the number-of-deposit estimates with the grade-and-tonnage model. For reasons of consistency, determining a mineral-deposit density requires unambiguous definitions of what is a deposit and what are the rules for delineation. Mineral deposit-density models start with the areas of well-explored control tracts where the number of deposits that are consistent with the grade-and-tonnage model are used to show variation of densities by deposit type.

Examples of the control tracts, areas of the tracts, and number of deposits within the tracts used for mineral deposit densities are shown in table 4.1. The examples of the information necessary for deposit-density models represent updated estimates of mineral deposit densities from various published sources (summarized in Singer et al., 2001; Singer, 2008). Densities for low-sulfide gold-quartz veins are discussed by Bliss, Menzie, Orris, and Page (1987), Bliss and Menzie (1993), and Lisitsin et al. (2007). These low-sulfide gold-quartz vein deposits are defined in the descriptive model by Berger (1986) and are consistent with the grade-and-tonnage model by Bliss (1986). It is important to note that the same proximity rule used to construct the grade-and-tonnage model was used to define deposits for the deposit densities; that is, workings within 1.6 km of each other were treated as part of the same deposit. Low-sulfide gold-quartz vein control areas, that is, explored permissive tracts (chapter 7) defined on the basis of standards from the descriptive model, consist of the presence of metavolcanic, metasedimentary, and ophiolitic rocks of accreted terranes. Plutonic rocks (within 40 km of veins) also intrude these terranes, and the metamorphic grade is greenschist facies or lower (Bliss, Orris, and Menzie, 1987). Reported densities for low-sulfide gold-quartz veins vary over a rather narrow range from 0.0033 to 0.0054 deposits/km^2.

Mineral deposit densities of three subtypes of volcanogenic manganese deposits (Mosier and Page, 1988) also rely on deposits that are consistent with their respective descriptive and grade-and-tonnage models. The examples in table 4.1 of mineral deposit densities for placer gold, kuroko massive sulfide (Mosier, Singer, and Berger, 2007), porphyry copper (Singer, Berger, Menzie, and Berger, 2005), Climax porphyry molybdenum, and wolframite-quartz vein deposits further demonstrate the variability of estimates and some consequences of assumptions about permissive areas and extent of exploration. Some of the assumptions are addressed below.

Permissive Area Sizes

One assumption that might be made is that mineral deposit densities are constant across all sized control tracts. A common way to estimate deposit

Table 4.1. Mineral deposit-density control areas.

Deposit Type[a]	Control Area Location[b]	Area (km^2)	Number	Median Tons (millions)	Total Tons (millions)	Source
Podiform Cr	Placer, CA	51	86	0.000065	0.0215	Singer (1994a)
Podiform Cr	Nevada, CA	39	38	0.000044	0.00740	Singer (1994a)
Podiform Cr	Sierra, CA	35	17	0.000061	0.00591	Singer (1994a)
Podiform Cr	Amador, CA	19	8	0.000120	0.00119	Singer (1994a)
Podiform Cr	Butte, CA	100	37	0.000085	0.0290	Singer (1994a)
Podiform Cr	El Dorado, CA	94	30	0.000136	0.142	Singer (1994a)
Podiform Cr	Calaveras, CA	66	17	0.000254	0.00765	Singer (1994a)
Podiform Cr	Fresno, CA	133	31	0.000121	0.0492	Singer (1994a)
Podiform Cr	Tehama, CA	151	33	0.000072	0.0116	Singer (1994a)
Podiform Cr	Tuolumne, CA	107	23	0.000190	0.0511	Singer (1994a)
Podiform Cr	Stanislaus, CA	38	6	0.000216	0.00333	Singer (1994a)
Podiform Cr	Plumas, CA	208	26	0.000029	0.00209	Singer (1994a)
Podiform Cr	St. Barbara, CA	50	6	0.000014	0.000195	Singer (1994a), (2008)
Podiform Cr	Tulare, CA	69	7	0.000242	0.00445	Singer (1994a)
Cuban Mn	Cuba	1,200	120	0.00293	7.97	Mosier and Page (1988)
Podiform Cr	San Louis Obispo, CA	187	16	0.000110	0.0572	Singer (1994a), (2008)
Podiform Cr	Glenn, CA	80	6	0.000540	0.0360	Singer (1994a)
Podiform Cr	Shasta, CA	261	19	0.000130	0.0217	Singer (1994a)
Podiform Cr	Siskiyou, CA	1,221	87	0.000046	0.0917	Singer (1994a)
Kuroko	West Shasta, CA	117	8	0.266	11.5	Mosier et al. (2007)
Podiform Cr	Del Norte, CA	711	47	0.000101	0.0570	Singer (1994a)
Podiform Cr	Alameda, CA	31	2	0.000260	0.00155	Singer (1994a), (2008)
Podiform Cr	Mendocino, CA	140	9	0.000021	0.00104	This study

(*continued*)

Table 4.1. (*continued*)

Deposit Type[a]	Control Area Location[b]	Area (km^2)	Number	Median Tons (millions)	Total Tons (millions)	Source
Podiform Cr	Humboldt, CA	271	16	0.000048	0.00399	Singer (1994a)
Podiform Cr	Santa Clara, CA	112	6	0.000135	0.00103	Singer (2008)
Podiform Cr	Lake, CA	277	14	0.000034	0.00232	Singer (1994a), (2008)
Podiform Cr	Monterey, CA	60	3	0.000017	0.000308	Singer (1994a), (2008)
Podiform Cr	Napa, CA	238	11	0.000137	0.00761	Singer (1994a)
Podiform Cr	Trinity, CA	1,034	46	0.000032	0.00646	Singer (1994a)
Kuroko	Jerome, AZ	24	1	29.000	29.0	Mosier et al. (2007)
Podiform Cr	Sonoma, CA	147	6	0.000331	0.00531	Singer (1994a), (2008)
Kuroko	Binghampton, AZ	26	1	0.363	0.363	Mosier et al. (2007)
Kuroko	Spain	1,300	48	2.046	1022	Mosier et al. (2007)
Kuroko	Chestatee, GA	27	1	1.100	1.10	Mosier et al. (2007)
Franciscan Mn	Hokaido, Japan	3,890	117	0.00293	2.04	Mosier and Page (1988)
Tungsten vein	SE China	140	4	0.55976	2.24	Singer et al (2001)
Kuroko	North Haven, ME	36	1	0.050	0.050	Mosier et al. (2007)
Kuroko	East Shasta, CA	73	2	0.306	0.771	Mosier et al. (2007)
Franciscan Mn	CA	23,700	450	0.00015	0.904	Mosier and Page (1988)
Cyprus m.	Cyprus	1,016	14	0.925	43.6	Mosier et al. (2007)
Kuroko	Castine Fm., ME	236	3	0.182	1.00	Mosier et al. (2007)
Kuroko	Ashland, AL	81	1	1.300	1.30	Mosier et al. (2007)
Kuroko	Kunitomi, Japan	416	5	2.100	2.10	Mosier et al. (2007)
Kuroko	Yavapai, AZ	89	1	1.429	1.43	Mosier et al. (2007)
Kuroko	Copper Hill, CA	424	4	0.315	1.82	Mosier et al. (2007)
Cyprus m.	Betts Cove, CNNF	433	4	2.147	20.3	Mosier et al. (2007)
Kuroko	Hillabee, AL-GA	218	2	0.497	2.84	Mosier et al. (2007)

Kuroko	Hokoruku, Japan	900	8	10.444	129	Singer et al (2001)
Kuroko	Dominican Republic	338	3	1.004	4.74	Mosier et al. (2007)
Kuroko	Quoddy Fm., ME	233	2	0.545	1.15	Mosier et al. (2007)
Kuroko	Standing Pond, VT	123	1	0.050	0.050	Mosier et al. (2007)
Kuroko	Pecos, NM	149	1	2.090	2.09	Mosier et al. (2007)
Cuban Mn	Fiji	2,720	18	0.00369	0.170	Mosier and Page (1988)
Placer gold	Kenai, AK	3,260	20	1.07152	21.4	Singer et al (2001)
Kuroko	Gopher Ridge, CA	1,343	8	0.140	3.67	Mosier et al. (2007)
Cyprus m.	Smartville, CA	518	3	0.128	1.51	Mosier et al. (2007)
Kuroko	Flin Flon-Snow Lake	2,656	15	0.809	79.2	Mosier et al. (2007)
Placer gold	Weisman, AK	2,760	15	1.07152	16.1	Singer et al (2001)
Kuroko	Tasmania	825	4	14.804	157	Mosier et al. (2007)
Kuroko	Kutcho Creek, CNBC	243	1	22.600	22.6	Mosier et al. (2007)
Kuroko	Buchans, CNNB	1,051	4	4.514	20.1	Mosier et al. (2007)
Kuroko	Hornblende Gneiss, GA	587	2	0.278	0.706	Mosier et al. (2007)
Kuroko	Hawley-Bernard, VT	297	1	0.902	0.90	Mosier et al. (2007)
Qtz.-gold	Bendigo, AUVT	7,000	23	0.427	60.9	Lisitsin et al. (2007)
Kuroko	Ammonoosuc, ME	1,136	3	0.136	0.650	Mosier et al. (2007)
Cyprus m.	Big Mike, NV	420	1	0.100	0.100	Mosier et al. (2007)
Cyprus m.	Lokken, Norway	949	2	1.581	25.1	Mosier et al. (2007)
Cyprus m.	Sunro, CNBC	480	1	2.780	2.78	Mosier et al. (2007)
Kuroko	Snake River, OR	1,071	2	4.469	39.5	Mosier et al. (2007)
Kuroko	Myra Falls, CNBC	1,117	2	1.176	5.45	Mosier et al. (2007)
Kuroko	Winterville Fm., ME	621	1	33.000	33.0	Mosier et al. (2007)
Porphyry Cu	Puerto Rico (8)	1,552	2	172.0	344	Singer et al. (2008)
Porphyry Cu	Eocene, Chile (12)	6,913	6	1,285.0	5,002	Cunningham et al. (2008)

(*continued*)

Table 4.1. *(continued)*

Deposit Type[a]	Control Area Location[b]	Area (km^2)	Number	Median Tons (millions)	Total Tons (millions)	Source
Porphyry Cu	Mio-Plio, AGNT (14d)	5,770	5	300.0	3,306	Cunningham et al. (2008)
Porphyry Cu	Olig., Chile (11)	2,429	2	553.0	1,638	Cunningham et al. (2008)
Porphyry Cu	Eoc-Olig., Chile (10)	25,690	16	1,852.9	85,581	Cunningham et al. (2008)
Kuroko	Orient, Cuba	3,390	2	2.391	11.9	Mosier et al. (2007)
Porphyry Cu	Mio-Plio, Chile (14b)	9,284	4	18,671.0	47,284	Cunningham et al. (2008)
Porphyry Cu	Mio Central, Peru (6)	53,186	27	518.4	22,745	Cunningham et al. (2008)
Porphyry Cu	Mio, Chile (14a)	21,721	8	186.5	5,383	Cunningham et al. (2008)
Kuroko	Rudny-Altai, Russia	20,539	8	1.242	206	Mosier et al. (2007)
Porphyry Cu	Eoc, Peru (9a)	30,154	11	233.1	6,070	Cunningham et al. (2008)
Porphyry Cu	Molong, AUNS (19)	11,500	4	133.0	1,592	Singer et al. (2008)
Porphyry Cu	Pale, Peru, Chile (8)	69,087	24	596.5	21,121	Cunningham et al. (2008)
Climax Mo	CO	12,000	4	240.1	960	Singer et al. (2001)
Porphyry Cu	Mio-Plio, Chile (14c)	24,048	8	657.0	7,025	Cunningham et al. (2008)
Porphyry Cu	West Philippine (1)	47,100	16	165.0	2,408	Singer et al. (2008)
Climax Mo	NM	3,200	1	423.9	424	Singer et al. (2001)
Porphyry Cu	Mio Colombia (5)	58,797	17	179.0	10,503	Cunningham et al. (2008)
Porphyry Cu	SE Eur. (15)	77,400	11	350.0	4,248	Singer et al. (2008)
Porphyry Cu	Jurassic Colombia (3)	67,709	17	384.2	12,423	Cunningham et al. (2008)
Kuroko	Urals, Russia	81,615	19	12.366	634	Mosier et al. (2007)
Porphyry Cu	Eoc, Colombia (1)	51,613	12	240.0	10,800	Cunningham et al. (2008)
Porphyry Cu	AZ– Mexico (7)	216,700	52	288.0	49,724	Singer et al. (2008)
Porphyry Cu	Permian AGTN (16)	29,080	6	308.6	3,553	Cunningham et al. (2008)
Porphyry Cu	Central Philippines (2)	76,580	13	78.0	2,627	Singer et al. (2008)
Porphyry Cu	Mio-Plio, Chile (13b)	41,799	8	2,945.0	13,965	Cunningham et al. (2008)

Porphyry Cu	Mio, Chile–AGTN (13a)	70,587	12	1,285.0	10,720	Cunningham et al. (2008)
Porphyry Cu	NV	32,800	5	159.0	1,986	Singer et al. (2008)
Porphyry Cu	Yulong, China (18)	93,400	8	92.2	1,698	Singer et al. (2008)
Porphyry Cu	Quesnellia, CNBC (5)	221,840	29	285.0	10,945	Singer et al. (2008)
Porphyry Cu	East Phillipines (3)	90,480	9	78.0	3,897	Singer et al. (2008)
Porphyry Cu	N. Sulawesi, INDO (4)	34,075	2	165.0	330	Singer et al. (2008)
Porphyry Cu	Kaz (16)	236,390	17	213.0	7,028	Singer et al. (2008)
Porphyry Cu	Cret. Peru (7)	107,297	6	75.3	2,738	Cunningham et al. (2008)
Porphyry Cu	Stikinia, CNBC–WA	255,700	15	171.0	4,379	Singer et al. (2008)
Porphyry Cu	East China (17)	146,320	6	314.0	3,098	Singer et al. (2008)
Porphyry Cu	Eoc, AGTN (15)	83,204	6	194.0	3,784	Cunningham et al. (2008)

[a] Podiform Cr = podiform chromite, Cuban Mn = Cuban volcanogenic manganese, Kuroko = kuroko massive sulfide, Franciscan Mn = Franciscan volcanogenic manganese, Cyprus m. = Cyprus massive sulfide, Qtz.-gold = low-sulfide quartz-gold vein, Climax Mo = Climax porphyry Mo.

[b] CA = U.S. California, AZ = U.S. Arizona, Spain = Iberian pyrite belt, Spain–Portugal, GA = U.S. Georgia, ME = U.S. Maine, AL = U.S. Alabama, CNNF = Canada Newfoundland, VT = U.S. Vermont, NM = U.S. New Mexico, AK = U.S. Alaska, Flin Flon-Snow Lake = Canada Manitoba-Saskatchewan, Tasmania = Australia Tasmania, CNBC = Canada British Columbia, CNNB = Canada New Brunswick, AUVT = Australia Victoria, NV = U.S. Nevada, OR = U.S. Oregon, CO = U.S. Colorado, WA = U.S. Washington, Mio-Plio = Miocene-Pliocene, Mio = Miocene, Pale = Paleocene, Eoc = Eocene, AGNT = Argentina, INDO = Indonesia, SE Eur = Carpathian-Balkan, Kaz = east-central Kazakhstan.

density is to average the number of deposits per permissive area over a number of permissive areas. This practice ignores possible effects of size of permissive area on deposit densities and ignores the frequency distribution and variability of densities. Each of these can have a significant effect on determining the best way to estimate uncertainty and reduce possible bias in estimated number of undiscovered mineral deposits. These issues are examined through studies of podiform chromite, porphyry copper, and volcanogenic massive sulfide deposit densities.

Density of Podiform Chromite Deposits

On the basis of number of discovered exposed podiform chromite deposits and area of permissive rocks, the deposit density, averaged over the twenty-eight sample areas in California (table 4.1), is 0.233 deposits/km^2 ultramafic rock (Singer, 1994a, 2007). This estimate is of questionable value, however, because the frequency distribution of the untransformed variable (deposits/area) is significantly skewed and peaked (figure 4.2). Thus, a few high values have a very large influence on the estimate, and probabilistic estimates of the number of undiscovered deposits are likely to be biased.

One way to make probabilistic estimates in this situation would be to use the mean and standard deviation of the transformed data and the normal distribution. The frequency distributions of the logarithms of the area of ultramafic rock, number of podiform chromite deposits, and number of podiform deposits per square kilometer of ultramafic rock are not significantly different from normal distributions. Note that it is more appropriate to model the distribution of number of deposits with a discrete distribution such as the negative binomial (Agterberg, 1977). As noted by Agterberg (1984), the continuous lognormal is equivalent to the discrete negative binomial, but use of the lognormal can lead to discrepancies for small frequencies.

On the basis of the lognormal distribution and the observed values, in 90 percent of the cases, the density of podiform chromite deposits would be ≥0.046 deposits/km^2, in 50 percent of the cases it would be ≥0.094 deposits/km^2, and in 10 percent of the cases it would be ≥0.538 deposits/km^2. The median estimate of 0.094 deposits/km^2 is nearly the same as an estimate made by dividing the total number of deposits (653) by the total area of ultramafic rock (5,930 km^2), that is, 0.110 deposits/km^2. This result suggests that the method of calculating the deposit density with untransformed data probably yields reliable estimates of the median density only when the permissive area is quite large. However, unless many areas are used, there is no way to estimate variability and, consequently, no way to make probabilistic estimates of the number of deposits without additional information. The value of multiple control areas and later the need to consider the size of permissive tracts are examined below for porphyry copper deposits.

Density of Porphyry Copper Deposits

Broad volcanic arcs that formed at approximately the same time are the fundamental unit for the delineation of belts permissive for porphyry copper deposits (Singer, Berger, Menzie, and Berger, 2005). Much of the information in that report is from the database of porphyry copper deposits of the world (Singer, Berger, and Moring, 2005) and various geologic maps and additional materials. For twenty-one porphyry copper settings from around the world, permissive belts were delineated and measured, and deposit densities were estimated. Porphyry copper belts were delineated and selected on the basis of three main features: (1) extensive exploration for porphyry copper deposits, (2) a definable geologic setting of the porphyry copper deposits in volcanic arc environments, and (3) a diversity of ages of porphyry copper deposit formation. The twenty-one belts in that report contain about 75 percent of the total amount of copper in all known porphyry copper deposits. In the study, the only deposits that are counted as porphyry copper deposits, rather than prospects, are those that have been drilled in three dimensions and have published estimates acceptable for the grade-and-tonnage models as reported in Singer, Berger, and Moring (2005). Porphyry copper prospects are counted separately here because in any subsequent resource assessment, only those undiscovered deposits that are represented by the grade-and-tonnage models are estimated—none, some, or all, of the known prospects might belong to that population.

Tracts containing porphyry copper deposits, referred to as control areas in Singer, Berger, Menzie, and Berger (2005), were outlined along borders of hosting geologic terrane and/or along regional zones of related porphyry intrusions with modifications on the basis of deposit ages and distributions of major structures. Eleven of the porphyry copper control areas were outlined on the basis of geologic maps of 1:1,000,000 to 1:1,500,000 scale; geologic maps of 1:2,000,000 to 1:2,500,000 scale were used for another eight control areas, and two belts were delineated at 1:240,000 to 1:250,000 scales.

The extent to which rocks or sediments younger than the porphyry copper deposits cover the permissive belts or deposits needs to be examined for two reasons. First, in most situations, covered permissive tracts cannot be considered to be well explored. Therefore, inclusion of covered parts of permissive tracts would improperly increase the size of the permissive tract. Second, counting mineral deposits that have been found under cover as part of the deposit density would distort density statistics and lead to misleading or biased estimates of how many undiscovered deposits might be present in an assessed tract. Boundaries of mapped rock units form the primary basis for drawing limits of permissive belts or tracts. Preliminary control area boundaries are typically extended using interpolated geology and geophysical surveys, such as aeromagnetics, to identify where younger rocks or sediments conceal permissive rocks. Scale of the maps can have a strong effect on the extent of cover portrayed—detailed maps commonly show more cover than do regional maps. Due to their large aerial extent, porphyry copper deposits located mostly under younger cover can have part of their alteration zones exposed. To prevent miscounting deposits that belong to the exposed permissive terrain, the percentage of cover associated with each deposit was estimated and density estimates were made by counting only those deposits that are at least 50 percent exposed. Thus, deposit-density estimates are formed from known deposits that are consistent with the grade-and-tonnage models and are located in exposed control areas. Of the 241 porphyry copper deposits that are in the twenty-one delineated control areas and that are consistent with the grade-and-tonnage models, 17 percent or forty-one deposits belong to the covered population.

Percent cover for each deposit was measured from detailed maps and cross sections of deposits. The area of each belt (in square kilometers) and the percentage of each belt covered by postmineral rocks, sediments and ice were estimated. Before deposit-density estimates were made, area permissive for a deposit type should be adjusted for the portion that is covered by removing the covered portion because it is considered poorly explored.

Estimates of porphyry copper deposit density (in deposit per square kilometer) were formed as the ratio of exposed deposit number to the total exposed belt area—covered explored deposits were not included in calculation. To make the deposit-density estimates accessible to more readers, we have also scaled the estimates by multiplying each by 100,000. These modified deposit densities can be considered as the number of porphyry copper deposits per 100,000 km^2 of permissive rock—this form of estimate may be more easily discussed and remembered because the estimates represent whole numbers of deposits. A histogram of porphyry copper densities per 100,000 km^2 using the data in table 4.1 has a skewed distribution like that observed for podiform chromite and volcanogenic massive sulfide deposits (figure 4.2). With skewed distributions such as in figure 4.2, a few high values have a very large influence on the mean density. However, the mean is only one measure of central tendency. Uncertainty is shown by the spread of the number-of-deposit estimates associated with the 90 percent to the 10 or 1 percent quantiles—with larger differences suggesting greater uncertainty. The estimates plotted in figure 4.2 provide the basis for probabilistic estimates of porphyry copper density that are not adversely influenced by a few large values. Ninety percent of the belts have densities of two or more deposits, 50 percent have densities of nine or more deposits, and 10 percent of the belts have densities of twenty-nine or more porphyry copper deposits per 100,000 km^2. Although these probabilistic estimates are an improvement over point estimates of density, they can be significantly improved by conditioning these guidelines by the size of permissive tract.

Permissive Areas

When samples are taken that represent small areas, there will be large differences from sample to sample according to sampling theory. In a number of well-explored regions that each represented a large area, one would expect much lower relative variability among the estimates. Thus, the scale of observations should have an effect on the variability of mineral deposit densities. For example, if tracts or belts representing larger regions were used to make a histogram of deposit density, one would expect that there would be fewer observations at the higher and lower densities, which is consistent with less variability of the estimates.

The relationship of sample size influencing deposit densities was recognized by Agterberg (1977) in his studies of volcanogenic massive sulfide deposits of the Abitibi region of Canada and was extensively discussed by Bliss and Menzie (1993) in terms of distributions and spatial correlations of several deposit types. These studies of frequency distributions and spatial correlations are typically concerned with variability within mineral deposit districts, whereas here we are concerned with variability among larger areas.

A further improvement in understanding the variability of density estimates and in increasing the precision of number-of-deposit estimates can be made by examining the relationship between the area permissive for a deposit type and the number of deposits in the control tracts (Singer, 1994a). A plot of the number of podiform chromite deposits versus the area permissive using data in table 4.1 is shown in figure 4.3, along with the linear regression line and 80 percent prediction intervals to the regression estimates. Estimates of the number of podiform chromite deposits can be made from figure 4.3 by using the logarithm of ultramafic rock area on the x-axis projected to the lower prediction interval for the 90 percent estimate of number of deposits, to the regression line for the 50 percent estimate, and to the upper prediction interval for the 10 percent estimate. A regression of the area of ultramafic rock on the number of known podiform chromite deposits was presented in Singer (1994a). This form of estimation is compared to deposit-density estimates derived from the same tracts discussed above with the data in table 4.1. The linear regression line and prediction intervals to the regression estimates can be made by using the logarithm of ultramafic rock area on the x-axis projected to the lower prediction interval for the 90 percent estimate of number of deposits, to the regression line for the 50 percent estimate, and to the upper prediction interval for the 10 percent estimate.

Although the correlation coefficient is not particularly high ($r = 0.50$), it and the associated regression slope (figure 4.3) are significantly different from zero at the 1 percent level. The slope of the regression line ($b = 0.47$) is also significantly different from one. A slope of 1.0 would mean that a doubling of permissive area would result in a doubling of the estimated number of deposits; that is, the ratio of number of deposits to the size of the permissive area would be independent of the size of the permissive area. Thus, if the slope equals 1.0, then the ratio of number of deposits to the size of the permissive area would provide an unbiased estimate for areas of any size. The fact that the slope is significantly less than 1 means the ratio of deposits to size of the permissive area is a biased estimator in many cases. This

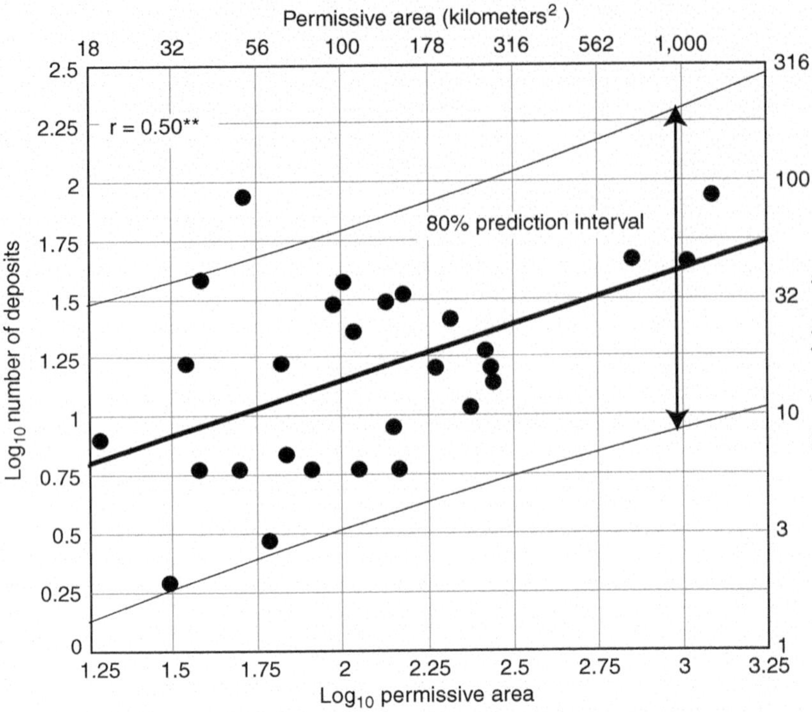

Figure 4.3 Podiform chromite permissive area versus number of deposits. Includes 80 percent prediction limits for deposits. Data are listed in table 4.1.

conclusion is reinforced by the observation that the correlation between deposits per area of ultramafic rock and areas of permissive ultramafic rock ($r = -0.55$) is significant at the 1 percent level; that is, the deposit density decreases as the size of the permissive area increases.

Porphyry copper deposit density is also inversely related to permissive area exposed. A similar negative relationship was found for the density of volcanogenic massive sulfide deposits (Mosier, Singer, and Berger, 2007). We are not certain of all of the reasons for this relationship to occur, but the relationship must be due in part to the rules used to define a sample that requires each tract to contain at least one known deposit. For large regions this works fine, but when the permissive tracts are quite small, packages of rock that appear to be permissive but do not contain a known deposit are excluded from the sample. Thus, a large permissive region that would be included as a permissive control tract would, when subdivided by more detailed and specific information, have some tracts that contain known deposits and some that would not be included in the sample because they do not contain a known deposit.

When assessing undiscovered mineral deposits by type or when developing control tracts, the base map selected is typically affected by there being only a limited number of geologic map scales of the region available. Additionally, publication scale of the assessment and time limits on the assessors affect map scales used. An effect of these limits is the common situation where the map scale selected is not ideal because the delineated permissive tract contains the geologic units that could host the deposits, but it may also include unreported units that could not contain the deposits (Singer and Menzie, 2008). It may also contain unreported geologic units that cover the geologic units of interest or unreported parts that are too deep to be permissive. In these situations, areas of delineated tracts are larger than necessary due to being inflated by nonpermissive areas that are not accounted for, or by covered areas that are poorly explored. More detailed information possibly allows identification of local settings that are not permissive for the deposit type. Therefore, groups of smaller tracts would tend to have higher deposit densities than the original larger tracts because of the exclusion of areas that appear permissive only at general scales but none of known deposits are excluded. These map scale effects certainly contribute to the negative relationships between deposit densities and permissive areas, as demonstrated by Singer and Menzie (2008). Another possible effect might be depth of emplacement of the deposit type. The argument is that deeper forming deposit types, such as orogenic gold, would be more abundant at greater depths due to the time required for exposure (Wilkinson and Kesler, 2007). However, this is really an issue of more permissive area existing at depth, not necessarily a different density of deposits.

Adjustments for Tract Size

Although we may not understand all of the reasons for the negative relationships between deposit densities and sizes of permissive tracts, we can use these relationships for predictive purposes. For porphyry copper deposits, the linear regression line and confidence limits to the regression estimates constructed with all thirty-three belts in table 4.1 are provided in figure 4.4. Most of the porphyry copper control areas in table 4.1 are from Singer, Berger, Menzie, and Berger (2005) and Singer, Berger, and Moring (2008), with updates for the South American control areas from Cunningham et al. (2007, 2008). The South American control tracts contain the total tract areas, the number of known deposits, and estimates of the number of undiscovered deposits. Estimates of the number of porphyry copper deposits can be made from figure 4.4 by using the permissive area on the x-axis projected to the lower prediction limit for the 90 percent estimate of number of deposits, to the regression line for the 50 percent estimate, and to the upper prediction limit for the 10 percent estimate. To be more precise, the two equations used for the prediction lines in figure 4.4 can be used:

$$R_{50} = -1.0252 + 0.42788 \log_{10}(\mathit{area}), \tag{4.1}$$

$$L_{90}, U_{10} = R_{50} \pm t\, s_{y|x} \sqrt{(1 + (1/n) + (\log_{10}(\mathit{area}) - 4.622)^2/(n-1)s_x^2)}, \tag{4.2}$$

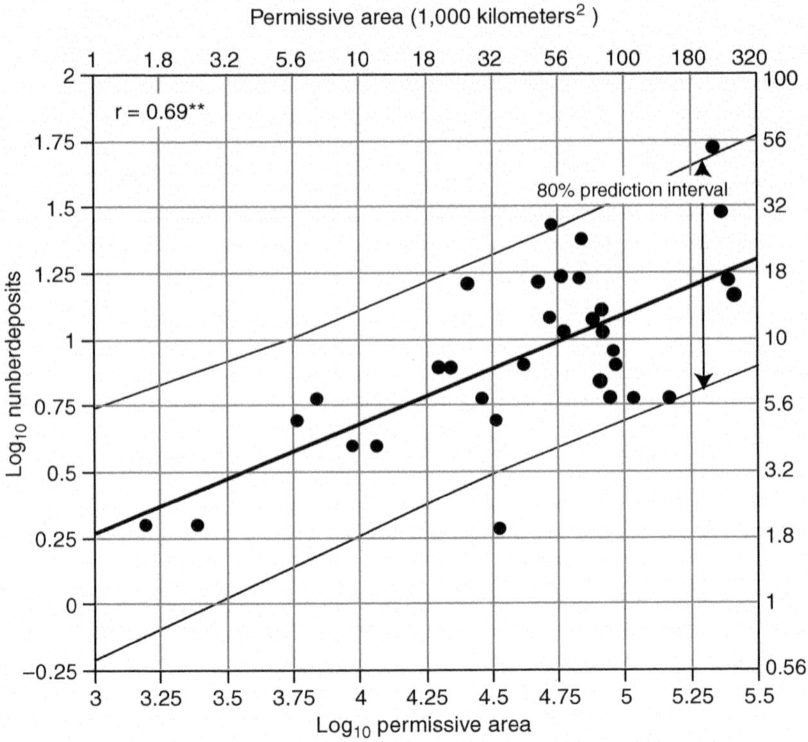

Figure 4.4 Porphyry Cu permissive area versus number of deposits. Includes 80-percent prediction limits for deposits. Data are listed in table 4.1.

where *area* is the permissive area in square kilometers, the mean area is 4.622 km^2, $t = t_{10,31df} = 1.309$, $s_{y|x} = 0.2444$, $n = 33$, $s_x^2 = 0.2912$, and the R_{50}, L_{90}, and U_{10} estimates need to be used as exponents to the power of 10. For example, if the permissive area is 25,000 km^2, then the 50th percentile estimate would be seven deposits (i.e., $10^{0.8567}$ or $10^{(-1.0252+0.42788\ \log10(25,000))}$). The 90th percentile estimate would be three deposits (i.e., $10^{(0.8567\ -\ 1.309\bullet 0.2444\ \surd(1+(1/33)+(4.398-4.622)\wedge 2\ /32\bullet 0.2912))}$), and the 10th percentile estimate would be fifteen deposits—these are approximately the same estimates as those determined from figure 4.4.

The expected number of deposits (1) can be estimated (Aitchison and Brown, 1963) as 10 to the power of

$$\log_{10} E(N) = \log_{10}(N_{50}) + (((\log_{10}(N_{10}) - \log_{10}(N_{50}))/t)^2)/2. \qquad (4.3)$$

The expected number of deposits estimated by the regression equation 4.3 can be used as an estimate of the expected number in some alternative guides (chapter 8) to the number of undiscovered deposits. Equation 4.3 can be used for porphyry copper deposits or any other deposit type for which the regression equations have been prepared, such as volcanogenic massive sulfide deposits. Estimates of the number of deposits made with density models are for the total number of deposits in a tract. In order to estimate the number of undiscovered deposits, the known number of deposits in a tract needs to be subtracted from the expected number (equation 4.3) to make a revised expected number of undiscovered deposits estimate. Calculation of probabilistic estimates of the number of undiscovered deposits after accounting for known deposits requires using the revised expected number and the variance based on the regression to estimate a new median and revised 90th and 10th percentile estimates. The variance from the regression can be estimated as

$$\text{var}_N = ((\log_{10}(N_{10}) - \log_{10}(N_{50}))/t)^2. \tag{4.4}$$

The revised estimate of the median number of deposits adjusted for the number of known deposits is estimated as 10 to the power of

$$\log_{10}(N_{50}) = \log_{10}(10^{E(N)} - \text{known number}) - \text{var}_N/2. \tag{4.5}$$

The value $\log_{10}(N_{50})$ can be used in place of R_{50} in equation 4.2 to make probabilistic estimates of the number of undiscovered deposits after accounting for the number of known deposits in a tract. The same procedures work for other deposit types, such as volcanogenic massive sulfide deposits discussed next.

Density of Volcanogenic Massive Sulfide Deposits

Control areas for volcanogenic massive sulfide deposits range in size from 24 km^2 to 82,000 km^2 and contain at least one exposed volcanogenic massive sulfide deposit (Mosier, Singer, and Berger, 2007). Typically, control area sizes represent the extent of exposed rocks permissive for volcanogenic massive sulfide deposits. As noted above, control areas with larger sizes may contain some nonpermissive rock units because of the way the units are generalized on the map, resulting in lower deposit-density values. Control areas with smaller sizes are more likely to contain only rock units permissive for volcanogenic massive sulfide deposits, thus resulting in higher deposit densities. Mineral deposit densities are inversely related to the size of the permissive control areas ($r = -0.8$, $n = 38$). This relationship suggests that the size of the permissive area can be used directly to estimate the number of deposits.

The linear regression line and prediction limits to estimate the number of deposits for individual permissive areas are based on thirty-eight control areas. For more precise estimates than can be shown in a plot, the following two equations are provided:

$$R_{50} = -0.5846 + 0.3846 \log_{10}(area), \quad (4.6)$$

$$L_{90}, U_{10} = R_{50} \pm t\, s_{y|x} \sqrt{(1+(1/n) + (\log_{10}(area) - 2.637)^2/(n-1)s_x^{\,2})}, \quad (4.7)$$

where *area* is the area that is permissive in square kilometers, the mean area is 2.637 km², t (Student's t at the 10 percent level with 36 degrees of freedom [$t_{10,36df}$]) is 1.306, $s_{y|x}$ (standard deviation of number of deposits given area) is 0.3379, n = 38, $s_x^{\,2}$ (variance of area) is 0.5258, and the R_{50}, L_{90}, and U_{10} estimates need to be used as exponents to the power of 10. For example, if the permissive area is 22,900 km², then the 50th percentile estimate would be twelve deposits (i.e., $10^{1.092}$ or $10^{(-0.5846 + 0.3846 \log10(22{,}900))}$). The 90th percentile estimate would be four deposits (i.e., $10^{(1.092 - 1.306 \cdot 0.3379 \sqrt{(1+(1/38) + (4.360 - 2.637)^{\wedge}2 /37 \cdot 0.5258)})}$), and the 10th percentile estimate would be thirty-five deposits. These estimates represent the total number of volcanogenic massive sulfide deposits in a permissive tract of 22,900 km², and any discovered deposits would need to be subtracted to estimate the number of undiscovered deposits. These estimates of 4, 12, and 35 or more deposits can be compared with those made by experts for a largely covered tract of the same area of 30, 65, and 85, or more kuroko massive sulfide deposits at the 90th, 50th, and 10th percentiles (tract LS04 in Ludington and Cox, 1996), demonstrating the value of this density method of estimating the number of deposits.

Densities of All Deposit Types

Analyses discussed above of podiform chromite, porphyry copper, and volcanogenic massive deposits strongly suggest that, at least within a deposit type, size of permissive tract should be used to help make unbiased estimates of number of deposits (figure 4.5). Examination of table 4.1 hints at another possible tool to help in making estimates—median deposit size seems to be related to deposit density. Examples in table 4.1 are ordered in decreasing density. Those familiar with some of these deposit types might notice that there is a general tendency of the deposits to be progressively larger down table 4.1. A plot of deposit size as measured by deposit tonnage versus deposit density (figure 4.6) demonstrates the strong relation between these variables across deposit types. The regression is significant at the 1 percent level, and deposit size explains 75 percent of the variability in density (Singer, 2008).

Rather than developing the relationship between deposit size and density, it seems that more progress might be made by considering both deposit size and permissive area to predict deposit density.

To be more precise, two equations estimated from the updated data in table 4.1 can be used (Singer and Kouda, 2008):

$$R_{50} = 4.2096 - 0.4987 \log_{10} area - 0.2252 \log_{10} size, \quad (4.8)$$

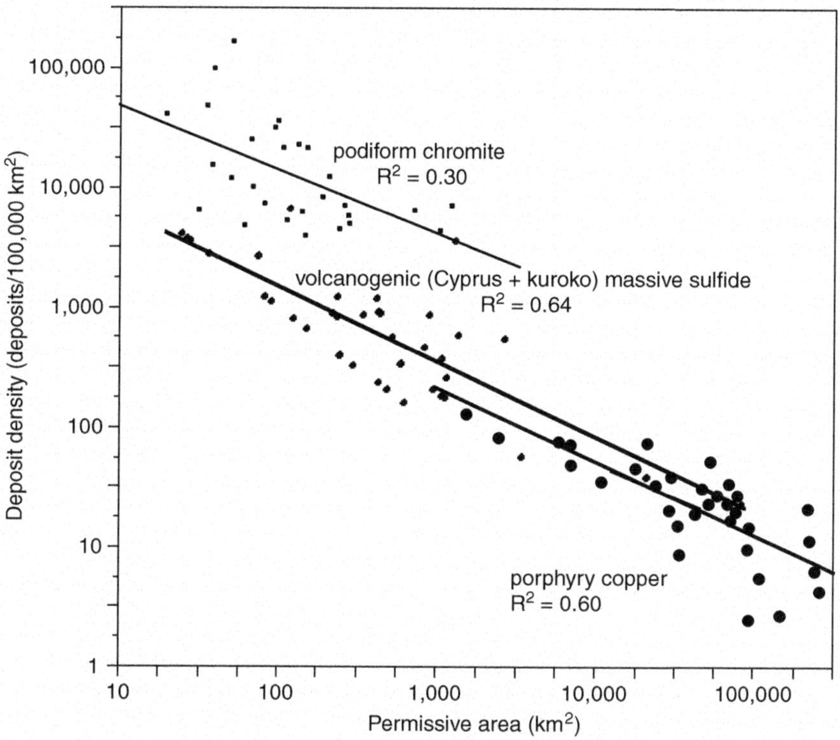

Figure 4.5 Density of deposits/100,000 km² versus the permissive area (in km²): podiform chromite deposits, volcanogenic (Cyprus + kuroko) massive sulfide deposits, and porphyry copper deposits with their respective regression lines. Densities of deposits/km² (without the 100,000 km² scaling used in the plot) are 1.64 *area*—0.53, 0.26 *area*—0.62, and 0.14 *area*—0.61 for chromite, sulfide, and copper, respectively. Data are listed in table 4.1.

$$L_{90}, U_{10} = R_{50} \pm t\, s_{y|s,a} \sqrt{}(1 + 1/n + (\log_{10} area - 3.173)^2 (\log_{10} size - -0.3292)^2/(n-1)s_s s_a), \quad (4.9)$$

where *area* is the permissive area in square kilometers, the mean area is 3.1727 km², the mean deposit *size* in log tonnage is −0.32923, $t = t_{10,106df} = 1.2896$, $s_{y|s,a} = 0.34841$, $n = 109$, $s_s = 2.6151$, $s_a = 1.1879$, and the R_{50}, L_{90}, and U_{10} estimates need to be used as exponents to the power of 10. For example, if the deposit type being assessed has a median size of 10 million tons and the permissive area is 900 km², then the 50th percentile estimate would be three deposits, that is, $(900/100{,}000) \bullet 10^{(4.2096 - 0.4987 \log 10(900) - 0.2252 \log 10(10))}$. The factor 900/100,000 is used to adjust for the regression equations estimating density/100,000 km², so we must account for the size of the tract (900 km²) to estimate the number of deposits. The 90th percentile estimate would be

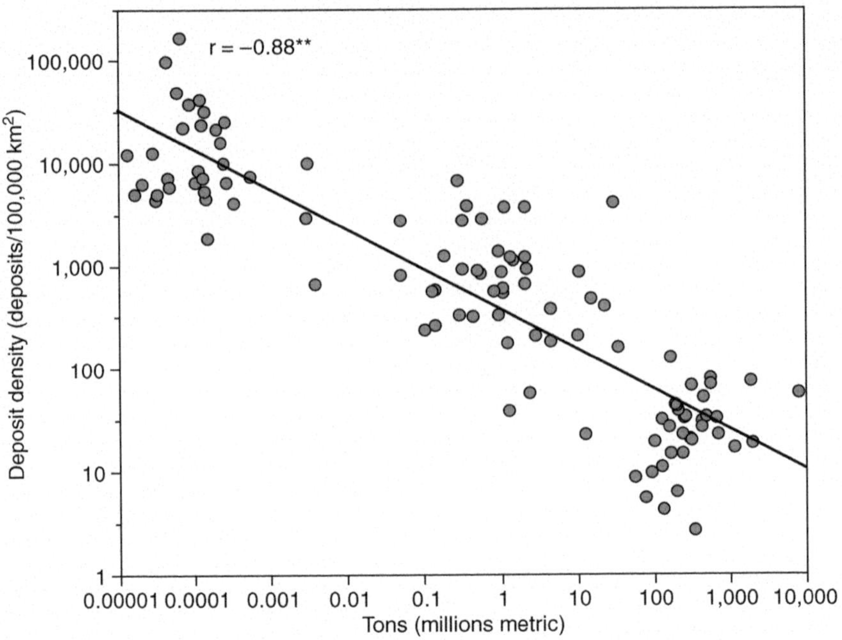

Figure 4.6 Median deposit size in millions of tons for all deposit types versus deposit density in deposits/100,000 km^2. Coefficient of determination (R^2) = 0.77. Data are listed in table 4.1.

one deposit, that is, $(900/100{,}000)\bullet 10^{(2.5112\ -\ 1.2896\bullet 0.3484\ \surd\ (1+(1/109)\ +\ (2.954\ -\ 3.1727)\wedge 2\ \bullet\ (1.0\ -\ -0.3292)\wedge 2\ /108\bullet 2.615\bullet\ 1.188))}$, and the 10th percentile estimate would be eight deposits. Regression equation 4.8 using both permissive area and deposit size explains 91 percent of the variability in deposit density. Although the equation was developed with only 109 control areas, the high percentage of the variation in the number of deposits explained by area and deposit size suggests that the equation and its associated prediction interval equations are quite robust. A plot of deposit density compared to deposit size and size of permissive tract (figure 4.7) demonstrates how these variables are related across deposit types.

A comparison of estimates from equations 4.8 and 4.9 to those of experts is available from a study on orogenic gold under cover in the Stawell zone of Victoria, Australia, reported by Lisitsin et al. (2009). The combined estimates by five experts of the number of undiscovered gold deposits under cover was 10, 33, and 60 deposits at the 90, 50, and 10 percent certainty levels, whereas the regression estimates are 11, 33, and 94. Except for less uncertainty expressed by the experts as shown by their 10 percent estimate being lower than the regression estimate, the estimates are not different. The experts did not know the regression estimates at the time they made their estimates.

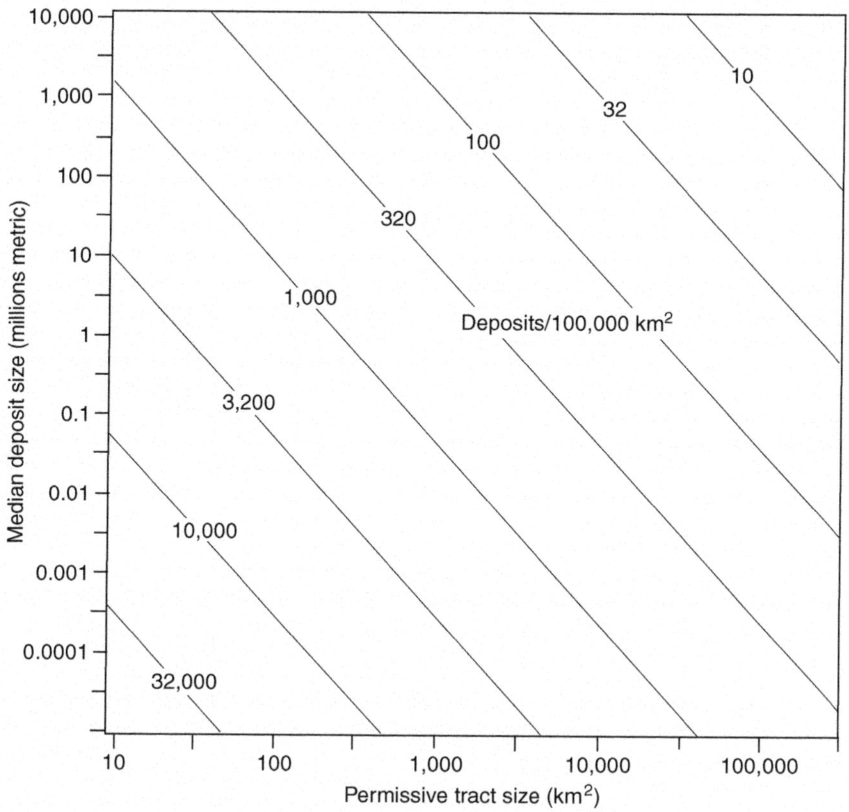

Figure 4.7 Variation of deposit density with change in size of permissive tract and median deposit size.

Because all available data were used to develop this density model, no truly independent verification with known well-explored deposits is possible at this time, but there are a few indirect tools to better understand the strengths and weaknesses of the model. The expected number estimate for the regression model can be used as the mean of a Poisson distribution in order to compare the widths of the 90th and 10th percentile estimates of the two procedures. In all cases examined, the Poisson distribution has a smaller range of estimates than does the regression model. Many subjective estimates made in various assessments seem to fall between the Poisson distribution and regression estimates. The differences between the regression estimates and the Poisson estimates are most noticeable at the number-of-deposit estimates associated with the 10th percentile. Without independent verification, it is not possible to determine which, if any, of these methods of making estimates to rely upon. However, if the goal is to mimic expert estimation, a close approximation can be made in many cases by using the universal regression model (equations 4.8 and 4.9) and substituting the upper, 10th percentile estimate by twice the median, 50th percentile estimate.

The same two variables also allow prediction of the total amount of mineralized rock of the deposit type in a permissive tract (Singer, 2008):

$$\text{Log}(\textit{total tonnage}) = -1.038 + 0.6784 \log(\textit{area}) + 0.6193 \log(\textit{size}), \tag{4.10}$$

where *total tonnage* is an estimate of the total tonnage of all mineralized rock in the delineated tract of the deposit size of *size*. About 95 percent of the variability in total tonnage in a tract is explained by this equation ($R^2 = 0.95$). Mineral deposits in this study range over more than ten orders of magnitude in size and ten different types of deposits providing a basis for generalization.

Concluding Remarks

The deposit-density models presented here are powerful tools in estimation of the number of undiscovered mineral deposits. The universal regression model presented here provides unbiased and reasonable estimates in most situations. These models were typically constructed with mineral deposits defined by certain consistently applied rules, such as distance to adjacent deposits; there are grade-and-tonnage models that follow these rules, and the permissive tracts or belts also were defined by consistent rules. As a consequence, these density models should not be expected to be applicable to other situations such as belts of rock that are not permissive. For most deposit types, the relationships developed here represent robust methods to estimate the numbers of deposits and the total resources in delineated tracts. For some deposit types, these predictors might not work properly because of the difficultly of delineating the boundaries of permissive rocks. For example, permissive tracts for Mississippi Valley Zn-Pb deposits might need to be delineated over broad areas where the only geologic information is the presence of carbonate rocks. A related situation might occur where there are widespread covering materials and the permissive geology under the cover represents a small part of the total area but cannot be separately delineated. In such cases, the regression equations presented here would tend to overestimate the resources. Careful integration of geophysics and extrapolated geology would reduce the number of such problems, however.

The strength of the relationships ($R^2 = 0.91$ for density and 0.95 for mineralized rock) argues for the broad use of these predictors of number of deposits and total resources. Of course, where specific deposit-density models exist, they are likely to lead to better estimates in that they would have lower variances and should be used rather than the general models presented here. Deposit densities can now be used to provide a guideline for expert judgment or used directly for estimates of the numbers of most kinds of mineral deposits. Even where deposit densities are applicable, they are probably best used as guidelines to final estimates. Density estimates should be the final estimates on the number of undiscovered deposits only where no information is available to refine the estimates.

5

Economic Analysis

Perspective

Estimated undiscovered mineral resources are based on grade-and-tonnage models made up of deposits of varying economic viability (chapter 6). Many deposits used in grade-and-tonnage models have not been developed because they cannot yet be mined economically. Although technological advances act over time to lower mining costs and environmental impacts, thereby allowing formerly uneconomic deposits to become operating mines, some deposits in these models might "never" be mined for one or more of a variety of reasons, including relatively low tonnages and grades, deep burial, or occurrence in or near environmentally sensitive areas.

Few nonacademic problems related to mineral resources are resolved by knowing the amount of metal that exists in some piece of land. Mineral policy issues and problems typically revolve around the effects of minerals that might be economically extracted. This is true if the problem concerns exploring or developing minerals, values of alternative uses of the land, or environmental consequences of minerals development. In resource assessments of undiscovered mineral deposits and in the early stages of exploration, including planning, a need for prefeasibility cost models exists. In exploration, these models separate economic from uneconomic deposits to help focus on targets that can benefit the exploration enterprise. In resource assessment, these cost models can be used to eliminate deposits that would probably be uneconomic even if discovered and allow estimation of the social value of the resources. Data used in grade-and-tonnage models do not necessarily represent economic deposits. Many of the deposits used in the models were found not to be economic and were not mined, whereas other deposits were mined long ago under economic conditions that no longer exist.

In this chapter we briefly explore some alternative measures of value used in assessments and then develop the basis for simplified economic filters to separate the clearly economic from the clearly uneconomic deposits. The equations used are not difficult, but they require care in their application because many of the apparently small cost factors can have large effects on the final economic discrimination.

Alternative Measures of Value

In addition to measuring or estimating resources in terms of the amount of metal, other measures have been advocated. One easy way to measure value is gross-in-place value, which can be calculated by multiplying the total amount of metal by some average price of the metal. The advantage of gross-in-place value is that it provides a common measure across different kinds of resources, such as minerals or timber. The disadvantage of gross-in-place value is that it neglects the costs of producing the materials, which could exceed the value of some or all of the undiscovered deposits, and it ignores the indirect benefits that flow from related economic activity. Even where some form of simplified cost model is accepted, the cost models may be used with regional input/output models to produce estimates of employment or taxes generated by possible mining of the yet-to-be-discovered resources and indirect economic benefits (Gunther, 1992). A variant of gross-in-place value applies the gross value only to the deposits that are predicted to be economic by a simplified cost filter (Harris and Rieber, 1993). In addition to measures such as net present value (used in this chapter), some analysts use measures such as internal rate of return and net smelter return (Wellmer, Dalheimer, and Wagner, 2008). The selection of an appropriate measure of value depends on the intended use of the assessment.

ASSESSMENT EXAMPLE

An assessment intended for general land use planning was of interest to some land managers near Medford, Oregon, in the 1980s because of the possible development of several nickel deposits. Existence of a nickel deposit near Riddle, Oregon, that was nearing the end of mining added to the interest in the other laterite Ni deposits in the area. At a time when the U.S. Geological Survey was not allowed to perform economic analyses, its assessment of mineral resources in the

(*continued*)

ASSESSMENT EXAMPLE (*continued*)

Medford Quadrangle (Singer et al., 1983) needed to provide useful information about the possibility of development of these Ni deposits. By plotting the average nickel grades and tonnage of ore of both the local deposits of interest and worldwide laterite Ni deposits, it was possible to show that the local deposits are much smaller and lower in Ni grades compared to known lateritic deposits elsewhere that were not being mined. The local deposits are significantly lower in grade and tonnage than other deposits that were being mined. While not an economic assessment, this analysis clearly demonstrates the relative disadvantages that the Oregon/California deposits would have to overcome if they were to be mined.

Simplified Cost Models

Here we examine simplified cost models for mining and beneficiation (milling). These models, which derived from those developed by the staff of the U.S. Bureau of Mines to assist in mineral-resource assessments (Camm, 1991, 1994; Smith, 1992), do not require an engineering background or detailed designs of full cost models to use. Thus, they can be used for estimating project economics in an assessment or the early stages of a mineral exploration program. The models can be applied to a number of types of deposits, and can be adjusted for changes in the location of the deposit or changes in prices. The models are capable of generating cost estimates at a level of uncertainty that is common to prefeasibility studies commonly performed by mining companies before extensive drilling of a prospect. Camm (1991) presented models of the capital and operating costs required to build and operate a mine and mill and the infrastructure that supports them. These models do not estimate the costs of preproduction exploration, permitting, environmental studies, taxes, corporate overhead, site reclamation, concentrate transportation, or smelter and refinery charges. Due to economics of scale, we can expect that the higher the rate of production per unit of time, the lower the cost per unit of volume. Thus, capacity to produce is a central focus of economic cost estimation. All cost estimates in the U.S. Bureau of Mines method are derived from the estimated daily mining capacity or its derived estimated mine life. Because of this, unbiased estimates of daily mining capacities are critical. Mining capacity and depth are used to estimate capital investments that vary by mining method and to estimate milling capital investments that vary by type of mill. Equations for operating costs of mines and mills are also a function of capacity and depend on the mining and mill method. These various estimates, along with assumed metal prices and rate of return, provide a basis for estimating an economic filter.

Capacity and Mine Life

In Camm's (1991) report, daily mining capacities are calculated using Taylor's rule (Taylor, 1978, 1986) from the total amount of ore in the deposit as

$$C_m = (T_d^{0.75})/71.725, \quad (5.1)$$

where C_m is mine capacity in metric tons per day, T is resource tonnage in metric tons, and 350 operating days per year (7 days/week) are assumed. Camm and Taylor used short tons, but we have chosen to keep our equations metric. For 260 operating days per year (5 days/week), the equation is

$$C_m = (T_d^{0.75})/53.282. \quad (5.2)$$

The general form of the cost models is

$$Y = A^*(C)^B, \quad (5.3)$$

where Y is the cost estimate, C is the daily capacity of the mine or mill, and A and B are constants. The capacity of the mine or mill varies depending on the tonnage of material being mined or milled and the rate at which the facility is operated. The daily capacity of the facility is the key variable in these models. Tonnage (T_d) is modified to account for recovery and dilution that vary from mining method to mining method and, to a lesser extent, mine to mine. The adjustment factors we use (table 5.1) are the same as Camm's (1991). The tonnage of material to be mined (T_m) is calculated from the deposit tonnage. This tonnage is adjusted for dilution and recovery by the following equation:

$$T_m = (T_d^0)^*(rf_m)^*(1+df_m), \quad (5.4)$$

Table 5.1. Mine dilution and recovery factors (Camm, 1991).

Mining Method	Dilution Factor (df_m), %	Recovery Factor (rf_m), %	Adjustment Factor $rf_m(1 + df_m)$
Open pit	5	90	0.945
Block caving	15	95	1.0925
Cut and fill	5	85	0.8925
Room and pillar	5	85[a]	0.8925
Shrinkage stope	10	90	0.99
Sublevel longhole	15	85	0.9775
Vertical crater retreat	10	90	0.99

[a] Assumes pillar extraction.

where T_d is the tonnage of ore, rf_m is the mine recovery factor, and df_m is the mine dilution factor. Table 5.1 lists these values for seven types of mines.

Mine life can be estimated from ore tonnage and daily capacity that was estimated in equations 5.1 and 5.2 as

$$L = T_d/(C_m \bullet 350), \tag{5.5}$$

assuming 350 operating days per year, or

$$L = T_d/(C_m \bullet 260), \tag{5.6}$$

assuming 260 operating days per year.

In addition, if the deposit is to be mined by open-pit methods, the tonnage of material to be mined must be adjusted to account for overburden. The stripping ratio, SR, of the deposit is the tonnage of waste material divided by the tonnage of ore and can be calculated as

$$SR = 0.667*(1.597*D*T_d^{-0.333} + 1)^3 - 1.667, \tag{5.7}$$

where T_d is the tonnage of ore in metric tons, and D is the depth to the bottom of the deposit measured in meters. The capacity, C_m, in tons per day, of an open-pit mine with strip ratio SR and tonnage of ore T_d may be calculated as

$$C_m + (SR + 1)*T_m/(L*dpy). \tag{5.8}$$

If the mine works 350 days per year (dpy) or 260 dpy, then equation 5.6 may be combined with equation 5.1 and rewritten, respectively, as follows:

$$C_m = ([SR + 1]*T_m^{0.75})/71.725 \tag{5.9}$$

$$C_m = ([SR + 1]*T_m^{0.75})/53.282 \tag{5.10}$$

The mining capacity estimates form the basis for estimating capital expenditures and operating expenses discussed below. However, milling capital and operating expenses in some cases are based on capacities that are different than mining capacities because the waste products from mining are not typically processed in the milling operations. For milling, the capacities are estimated as

$$C_{ml} = (T_m^{0.75})/71.725, \tag{5.11}$$

$$C_{ml} = (T_m^{0.75})/53.282. \tag{5.12}$$

Capital Expenditures

The capital and operating cost equations presented in this section (tables 5.2, 5.3) and the following section can be used if the deposits being evaluated are located in the western United States, on which these models were based. Outside of the western United States, cost-factor differences for individual cost categories can be used to modify base case equations (Long and Singer, 2001). For example, capital costs and operating costs in Alaska are estimated to be 1.80 and 1.40 times higher, respectively, than the base case. Analysis by Penney et al. (2004) demonstrated the importance of including infrastructure costs in considering the economics in remote locations (see chapter 10). To apply these models to Alaska, one must multiply the base case cost equations for appropriate categories by the associated cost differential to obtain a new cost model.

Table 5.2. Mine cost equations (Camm, 1991, except where modified).

Mining Method	Capital Cost	Operating Cost
Open pit	$2{,}920\ C_m^{0.917}$	$5.58\ C_m^{-0.148}$
Block caving	$70{,}000\ C_m^{0.790}$	$52.2\ C_m^{-0.217}$
Cut and fill	$1{,}310{,}000\ C_m^{0.461}$	$299\ C_m^{-0.294}$
Room and pillar	$104{,}000\ C_m^{0.644}$	$38.5\ C_m^{-0.171}$
Shrinkage stope	$190{,}000\ C_m^{0.620}$	$81.8\ C_m^{-0.100}$
Sublevel longhole	$121{,}000\ C_m^{0.552}$	$45.4\ C_m^{-0.181}$
Vertical crater retreat	$48{,}600\ C_m^{0.747}$	$55.1\ C_m^{-0.206}$

C_m = capacity in metric tons/day

Table 5.3. Mill cost equations (Camm, 1991, except where modified).

Milling Method	Capital Cost	Operating cost
Autoclave-CIL-EW	$104{,}000\ C_{ml}^{0.778}$	$84.5\ C_{ml}^{-0.196}$
CIL-EW	$53{,}800\ C_{ml}^{0.745}$	$90.3\ C_{ml}^{-0.281}$
CIP-EW	$392{,}000\ C_{ml}^{0.540}$	$112.0\ C_{ml}^{-0.303}$
CCD-MC	$438{,}000\ C_{ml}^{0.584}$	$137\ C_{ml}^{-0.300}$
Float-roast-leach	$508{,}000\ C_{ml}^{0.552}$	$109\ C_{ml}^{-0.246}$
Flotation 1 product	$98{,}800\ C_{ml}^{0.667}$	$55.1\ C_{ml}^{-0.206}$
Flotation 2 product	$88{,}300\ C_{ml}^{0.702}$	$159\ C_{ml}^{-0.336}$
Flotation 3 product	$89{,}600\ C_{ml}^{0.708}$	$163\ C_{ml}^{-0.344}$
Gravity	$143{,}000\ C_{ml}^{0.529}$	$72.1\ C_{ml}^{-0.364}$
Heap leach	$312{,}000\ C_{ml}^{0.512}$	$34\ C_{ml}^{-0.223}$
Solvent extraction-EW	$15{,}500\ C_{ml}^{0.596}$	$3.26\ C_{ml}^{-0.145}$

C_{ml} = capacity in metric tons/day For abbreviations, see text.

Capital expenditures used here represent the total reported over the life of a mine. Frequently, mining operations are observed to spread out their capital expenditures by means of mine or mill expansions over a period of years. The simplified nature of the economic analysis used assumes that capital expenditures are made at the beginning of the first year and that mining/mill capacities remain constant until the deposit is depleted. Although it is not difficult to adjust by means of discounting any capital expenditures made after the initial expenditure so that all investments can be counted the same, it does involve some complexity to account for present values of variable rates of production that result from these investments.

The mill or beneficiating cost models have the same general form as those for mine cost, and like the mine cost models, the daily capacity of the facility is the key variable in these equations. The simplified cost models estimate capital and operating costs for eleven types of mills—carbon-in-leach-electrowinning (CIL-EW), autoclave-CIL-EW, carbon-in-pulp (CIP), countercurrent decantation-Merrill Crowe (CCD-MC), float-roast-leach, one-product floatation, two-product floatation, three-product floatation, gravity, heap leach, and solvent extraction-electrowinning. For each type of mill, Camm (1991) presented equations that estimate the capital and operating costs associated with up to ten categories of expenses, including labor equipment, steel, fuel, lube, tires, construction material, electricity, reagents, and sales tax. He also presented summary equations that estimate capital and operating costs directly. Table 5.3 lists his summary equations (modified to metric units) that estimate the capital and operating costs of each milling method.

Each cost estimate of underground mining methods assumes that the mine has an adit entry. Mines having a shaft entry need to have the additional capital cost of the shaft. The capital cost equation for a shaft after Camm (1991) is

$$\text{Shaft cost} = 322 \cdot D^{0.851} \cdot C_m^{0.489}, \tag{5.13}$$

where C_m is capacity of mine in metric tons per day, and D is depth of the shaft to the bottom of ore in meters.

Operating Costs

Operating cost of mines and mills (tables 5.2, 5.3) depend on the mining and milling method and are a function of capacity. Mines having a shaft entry need to have the additional operating cost of the shaft added. The operating cost equation for a shaft from Camm (1991) is:

$$\text{Shaft operating cost} = 0.838 \cdot D^{0.705} \cdot C_m^{-0.338}, \tag{5.14}$$

where C_m is capacity of mine in metric tons per day, and D is depth of the shaft to the bottom of ore in meters.

These simplified capital and operating cost models can be used as economic filters during the early stages of mineral exploration and in mineral-resource assessment.

Economic Filter

Given an appropriate mining method and depth of a deposit, the deposit's tonnage is all that is needed to estimate various mining costs using the equations above. The deposit's grade(s) can, when combined with assumed recoveries and metal prices, be used to estimate the deposit's ore value per ton. Value of production per year can be calculated by multiplying the difference between value per ton and total cost per ton by capacity per day times number of operating day per year (350 days used here).

The life of the mine estimate is then used with the value of production per year and an acceptable rate of return (10 percent used here) in a standard net present-value equation in a spreadsheet to estimate a deposit's present value of production. The present value of production minus the estimated capital expenditure for the deposit is the present value of the deposit. If the deposit's net present value is positive, the filter is predicting that the mine is profitable. Negative present values predict economic failure at the assumed metal prices and rate of return.

For a particular tonnage, the dividing (or breakeven) line between economic and uneconomic can be estimated by adding the estimated operating cost to the capital expenditure divided by capacity times operating days per year times the present value of a dollar for the life of the mine. That is,

$$BE = TOC + MOC/(350 \bullet C_{ml} \bullet PV), \quad (5.15)$$

where BE is the breakeven value ($/t), TOC is total operating cost ($/t), MOC is the total capital expenditure ($), 350 is days per year, C_{ml} is the mill capacity (t/day), and PV is the present value of one dollar at the selected rate of return for the life of the mine in years (Singer, Menzie, and Long, 1998). The breakeven value could be viewed as the grade (expressed in $/ton) at which the specific deposit and mining method are just economic. To account for variability and uncertainty in most of the inputs to these estimates, we have taken 0.7 and 1.3 of this breakeven value to estimate boundaries for uneconomic, marginal, and economic deposits.

As an example, the seventy-five deposits used in the polymetallic vein grade-and-tonnage model (Bliss and Cox, 1986) are assumed to be located 500 m below the surface and are mined by cut-and-fill method. The prices metals are copper, $3.40/lb; gold, $630/oz; silver, $11/oz; zinc, $1.50/lb; and lead, $0.50/lb. Using the equations above, 260 days per year mining, and a 10 percent return on capital, most of the deposits have negative net present values and are not

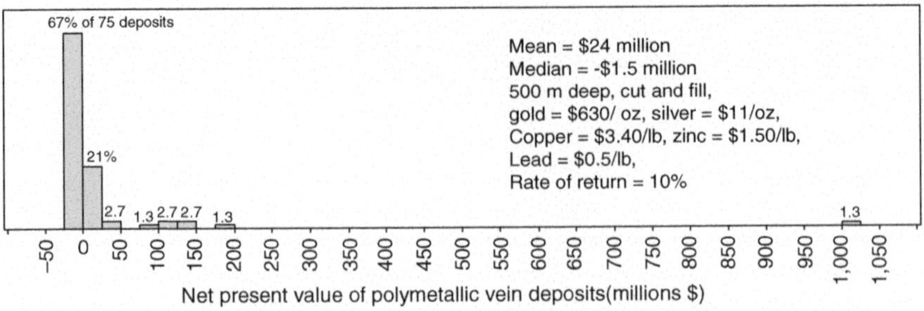

Figure 5.1 Histogram of the net present value of polymetallic vein deposits used in published grade-and-tonnage model (Bliss and Cox, 1986) and costs from Singer, Menzie, and Long, (2000).

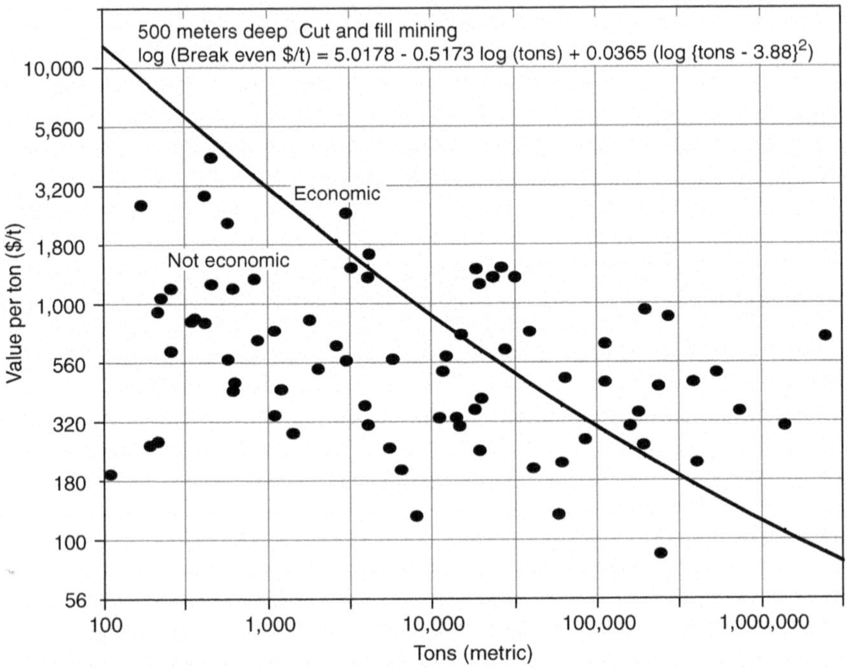

Figure 5.2 Breakeven line separating economic from uneconomic polymetallic vein deposits (same deposits and conditions as in figure 5.1).

economic to mine (figure 5.1). Despite the average net present value of these deposits being $24 million under the stated conditions, 67 percent of the deposits have negative values. Some example grade, tonnage, and economic costs and returns (tables 5.4, 5.5) show that the costs of mining smaller deposits exceed their values. This is further emphasized in figure 5.2, where even values greater than $1,000/ton do not guarantee economic success.

Table 5.4. Example of grades, tonnages, life, and capacity of 10 polymetallic vein deposits.

Tons (metric)	Cu (%)	Zinc (%)	Lead (%)	Au (g/t)	Ag (g/t)	Life (years)	Capacity (t/day)
1,100	0.5	4	12	0	230	1.3	3.3
7,990	0	2.98	1.85	0.016	65.8	2.1	15.0
212	0.12	0	6.1	7.2	277	0.9	1.0
18,400	0.026	5.89	25.2	0.032	3660	2.6	27.0
12,000	0	0	5	0	2060	2.3	20.0
410	11	0	1.4	21	6630	1.0	1.6
4,150	0	16.9	41	0.06	2628	1.8	9.0
5,490	0.29	4.02	5	2.9	29.2	1.9	11.0
2,600	0.29	6.4	17	12	324	1.6	6.3
195,000	0.23	18	16	9.2	290	4.7	160.0

Table 5.5. Estimated economic costs and returns for the 10 deposits in table 5.4.

Shaft Capital (thousand $)	Shaft Operating Cost ($/t)	Mine Capital (thousand $)	Mine Operating Cost ($/t)	Mill Capital (thousand $)	Mill Operating ($/t)	Present Value (million $)	Break even ($/t)	Value Ore ($/t)
114	45	2,270	210	204	106	−2.6	2,980	333
236	27	4,500	140	579	65	−6.0	1,000	124
62	22	1,280	300	86	160	−1.5	7,910	259
320	22	6,010	110	898	52	11.6	653	1,400
274	24	5,180	120	717	58	−2.2	809	596
80	58	1,610	260	121	137	−0.9	5,320	2,580
186	32	3,590	160	410	76	0.7	1,420	1,600
206	30	3,960	150	475	71	−4.7	1,220	243
157	36	3,050	170	320	86	−2.7	1,830	659
761	12	13,600	67	3,110	29	107.0	225	936

Summary

Although not all costs are included and the estimates are rough, these models serve to discriminate clearly uneconomic from clearly economic deposits at an early assessment stage. The forms of these equations appear to be robust, but the coefficients to the equations are based on costs estimated more than twenty years ago and need to be estimated using modern costs. In addition, the coefficients to the fundamental equations relating mine capacity to tons of ore (equations 5.1, 5.2) recently have been found to require new coefficients. Long's (2009) analysis shows that real mining capacities are higher than suggested by the Taylor equation and are significantly different for underground versus open-pit or underground block caving mining. These results in turn require reexamination of the relations between capacities and capital and operating costs. We hope this important research will be conducted in a timely manner.

6

Grade-and-Tonnage Models

Perspective

Mineral deposit models are important in quantitative resource assessments for two reasons: (1) grades and tonnages of most deposit types are significantly different (Singer, Cox, and Drew, 1975; Singer and Kouda, 2003), and (2) deposit types occur in different geologic settings that can be identified from geologic maps. If assessments were only conducted to estimate amounts of undiscovered metals, we would need contained metal models, but determining whether the metals might be economic to recover is an important quality of most assessments, and grades and tonnages are necessary to estimate economic viability of mineral deposits (see chapter 5). In this chapter, we focus on the first part of three-part assessments: grade-and-tonnage models. Too few thoroughly explored mineral deposits are available in most areas being assessed for reliable identification of the important geoscience variables or for robust estimation of undiscovered deposits, so we need mineral-deposit models that are generalized. Well-designed and well-constructed grade-and-tonnage models allow mineral economists to determine the possible economic viability of the resources in the region and provide the foundation for planning. Thus, mineral deposit models play the central role in transforming geoscience information to a form useful to policy-makers. Grade-and-tonnage models are fundamental in the development of other kinds of models such as deposit-density and economic filters.

Frequency distributions of tonnages and average grades of well-explored deposits of each type are employed as models for grades and tonnages of undiscovered deposits of the same type in geologically similar settings. Grade-and-tonnage models (Cox and Singer, 1986; Mosier and Page, 1988; Bliss, 1992a, 1992b; Cox et al., 2003; Singer, Berger, and Moring, 2008) combined with estimates of number of undiscovered deposits are the fundamental means of translating geologists' resource assessments into a language that decision-makers can use. For example, creation of a grade-and-tonnage model for rhyolite-hosted Sn deposits in 1986 demonstrated for the first time that 90 percent of such deposits contain less than 4,200 tons of ore. This made it clear that an ongoing research project by the U.S. Geological Survey on this deposit type could have no effect on domestic supplies of tin, and the project was canceled.

Grade-and-tonnage models specifically prepared for assessments show the frequencies of different sizes and grades of each mineral deposit type based on data collected on thousands of well-explored deposits from around the world. For each deposit type, these models help define a deposit, as opposed to a mineral occurrence or a weak manifestation of an ore-forming process. Data utilized to construct these models include average grades of each metal or mineral commodity of possible economic interest and the associated tonnage based on the total production, reserves, and resources at the lowest possible cutoff grade (Singer et al., 1980; Orris and Bliss, 1985; Bliss and Jones, 1988; Singer, Mosier, and Menzie, 1993; Cox et al., 2003; Singer, Berger, and Moring, 2008; Mosier, Berger, and Singer, 2009). These data represent an estimate of the endowment of each of many known deposits so that the final models can accurately represent the endowment of all undiscovered deposits (Singer, 1994b).

Representation of Grade-and-Tonnage Models

Grade-and-tonnage models are presented in a graphical format to make it easy to compare deposit types and to display the data. The plots show grade or tonnage on the horizontal axis, and the vertical axis always is the cumulative proportion of deposits (figures 6.1, 6.2, 6.3). All tonnages are reported in millions of metric tons. Grades not available (always for by-products) are represented as below the lowest value plotted and are therefore not shown. Plots of the same commodity or tonnages are presented on the same scale; a logarithmic scale is used for tonnage and most grades. Each dot represents an individual deposit (or, rarely, a district for some models), cumulated in ascending grade or tonnage. Where a large number of deposits is plotted (e.g., in Cox and Singer, 1986), individual numbers represent a count of the number of deposits.

Grade-and-tonnage data are usually displayed either as univariate or as bivariate plots. In univariate plots, the data are sorted from smallest to largest and are plotted against the proportion of the deposits that are as large as or larger than each deposit (figures 6.1, 6.2, 6.3). The median of the data (50th percentile) is either the observed median or is estimated from logged data,

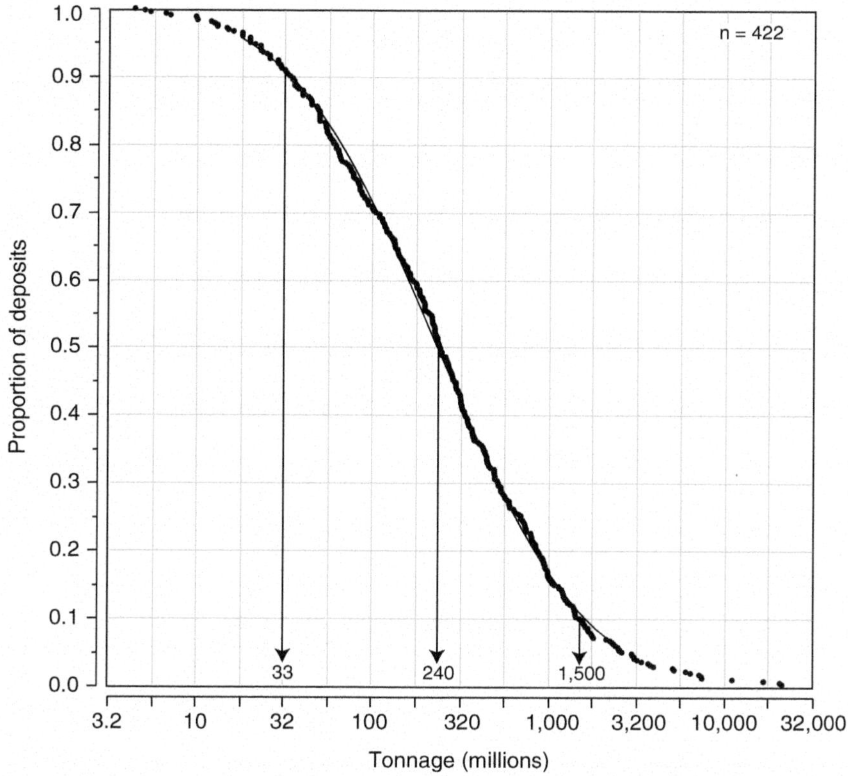

Figure 6.1 Cumulated frequency of ore tonnages of porphyry Cu deposits. Each dot represents an individual deposit. Intercepts for the observed 90th, 50th, and 10th percentiles and the lognormal distribution are plotted. Reprinted from Singer, Berger, and Moring (2008).

the 90th and 10th quantiles are either observed quantiles or are calculated using the standard deviation of the logged data, and a curve is fitted to these data. Observed quantiles are desirable when a lognormal distribution is not appropriate or when there are missing grades because fewer assumptions are required about why some grades are missing. The horizontal scale in each of these figures is logarithmic. A univariate plot is made for tonnage and each grade for which a significant proportion of the deposits report grades. The scales and intervals for tonnage and each commodity, such as copper, are the same for all deposit types to allow direct comparisons of types.

To compare amounts and qualities of resources among deposit types in one figure, deposit models may be plotted in grade-and-tonnage space (figure 6.4). Because data in grade-and-tonnage models vary logarithmically, plotting all of the deposits in several models may show so much scatter that comparison of central tendencies of the models may be lost. Therefore, it is common to plot an ellipse centered on the means of grade and tonnage plus

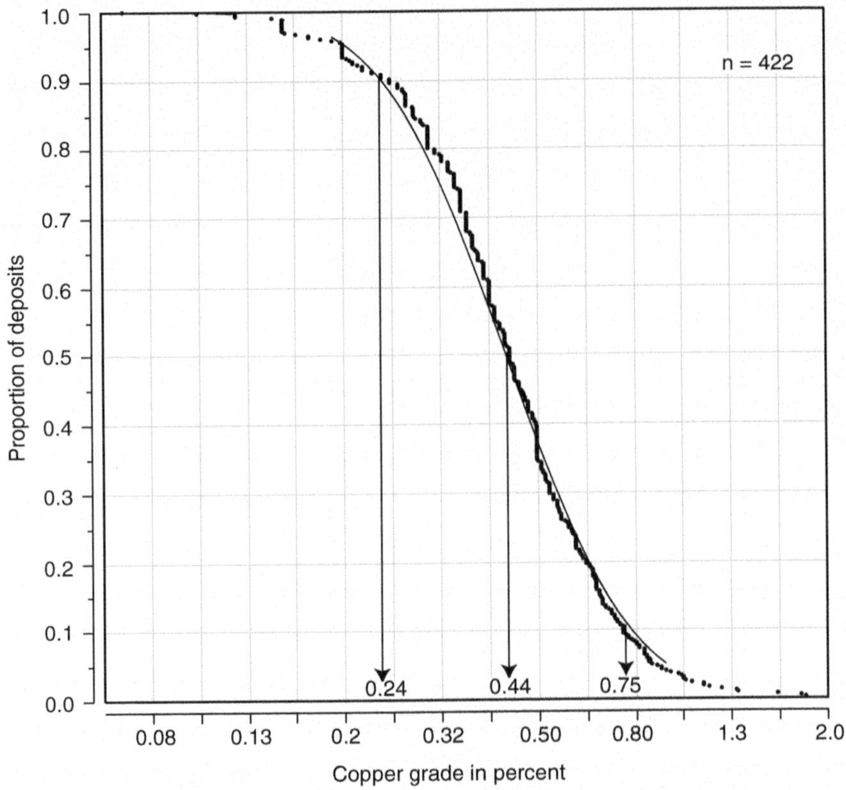

Figure 6.2 Cumulated frequency of copper grade of porphyry Cu deposits. Each dot represents an individual deposit. Intercepts for the observed 90th, 50th, and 10th percentiles and the lognormal distribution are plotted. Reprinted from Singer, Berger, and Moring (2008).

and minus one standard deviation and the long axis of the ellipse oriented relative to the correlation of grade and tonnage. Calculations of the mean and standard deviation are with logged data, and the plotted axes use a log interval. Centers of the ellipses represent medians because means of lognormal distributions are estimates of medians when the antilog is taken. Slopes of ellipses are different than zero only when the correlation between grade and tonnage is significantly different than zero. Each ellipse contains about 45 percent of the deposits of the type. To show effects of large deposits, an elephant is located at the median tonnage and grade of the five largest deposits (in terms of contained metal). Diagonal lines in figure 6.4 represent equal gold content. Points on a line all contain the same amount of gold, but as one moves to the right on the line, the grade at each point declines. Although they plot in different places in figure 6.4, many hot-spring Au-Ag deposits are similar to Comstock epithermal Au-Ag deposits in terms of their geologic characteristics. Hot-spring Au-Ag deposits are thought to have formed

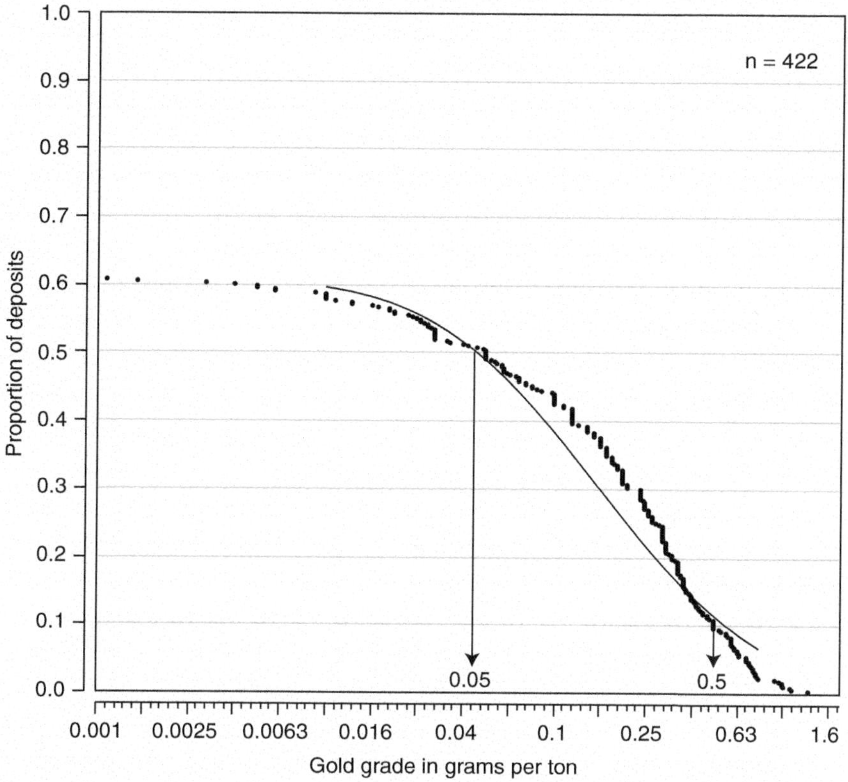

Figure 6.3 Cumulated frequency of gold grade of porphyry Cu deposits. Each dot represents an individual deposit. Intercepts for the observed 90th and 50th percentiles and the lognormal distribution are plotted. Reprinted from Singer, Berger, and Moring (2008).

in the upper parts of geothermal systems, whereas Comstock deposits are thought to form slightly deeper in these systems. The Witwatersrand districts of South Africa that have dominated world gold production for almost 100 years plot in separate grade-and-tonnage space than other gold-bearing deposits. Finally, notice the point marked "Bre-X," which represents the Busang deposit in Kalimantan, Indonesia. This is the grade and tonnage reported for one of the most famous mining scams in recent years. The fact that no deposit types fall anywhere near it demonstrates that grade-and-tonnage plots sometimes can be used to identify anomalous deposits.

When Is a New Model Needed?

The purpose of grade-and-tonnage models is to provide unbiased representations of the grades and tonnages of undiscovered mineral deposits in a tract or

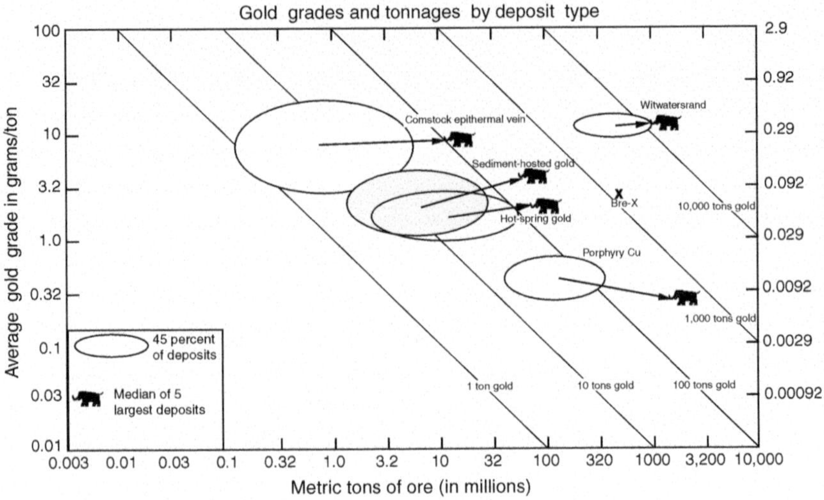

Figure 6.4 Average gold grades versus tonnages of some types of gold-bearing deposits. The X marking the grade and tonnage of the Busang deposit (company Bre-X) stands out.

belt. A new model is required in any situation where an existing grade-and-tonnage model can be shown to be a biased model of the undiscovered deposits. When only one or two explored examples of a deposit type are known in a local area, it is common to believe that the deposits represent a special subtype or new type because the deposits are almost never exactly the same as the "typical" deposit in every respect. Local deposit grades and tonnages are never the same as model median or mean grades or tonnages, and typically some unusual mineral or some trace element is present that is not mentioned in the model.

To avoid the situation where every deposit is considered to be unique and therefore prediction is not possible, the well-explored—that is, completely drilled—known deposits in the local area should always be tested to see if they are statistically different from the general model. The known deposits falling on the published grade-and-tonnage model is not an adequate test because all of the known deposits might have tonnages that are on the low tonnage part of the model, suggesting a mismatch between known deposits and the model. Tonnages and grades of the known deposits in an area being assessed should always be properly tested against the model before the assessment. With respect to the issue of providing representations of the grades and tonnages of undiscovered mineral deposits in a tract or belt, the most appropriate test is a *t*-test of the average tonnage and grade(s) of the local deposits compared to the model. Of course, all of the data should be in log form. The *t*-test gives the probability that the difference between the two means happened by chance alone. Typically, the tests have been if the probability of *t* is less than 0.01, then the difference is deemed significant, that is, not caused by chance, and we reject the hypothesis that the known deposits in the tract are a random sample from the model. The probability level of the test, 0.01, was selected in an attempt to balance the costs associated with making an error of rejecting a model when it is appropriate and the error of accepting a model when it is not appropriate (type I and type II errors). For each tract that contains at least one deposit that is deemed well enough explored to be included in the grade-and-tonnage model, the tonnages and grades of all of the well-explored deposits in that tract should be compared to the remaining well-explored deposits in the world that are part of the general grade-and-tonnage model. In situations where no well-explored deposits are known in a tract, we assume that the general model containing all deposits is the best representative of the undiscovered deposits in the tract, because we have no basis of selecting a more specific model for the tract. The cost of rejecting the general model, if in fact it is the correct model, is the introduction of a biased model. Rejection of the general model suggests that one is confident that the undiscovered local deposits contain less variability than does the general model. This is a strong statement that needs to be documented.

If the well-explored deposits are significantly different in size or grade, then the local deposits should be examined to see if they belong to a geologically homogeneous subset of the original grade-and-tonnage model. Only if all of these conditions are met should a new model be constructed along with a consistent descriptive model (Menzie and Singer, 1993). Where a different model is needed, the guiding principle should be to provide an unbiased model of the grades and tonnages of the undiscovered deposits in the tract. Given two models that provide unbiased estimates, pick the one with the larger number of deposits because it will be more robust.

AN EXAMPLE

Test of appropriateness of the global porphyry copper grade-and-tonnage model containing 370 (380 minus the 10 in Yulong) deposits to the Yulong belt of China's ten well-explored deposits in 2007:

$$t=(8.005-8.234)/0.2.21=-1.67$$

with 378 degrees of freedom, mean tonnage of Yulong deposits = 8.005, mean tonnage of the world = 8.234, and pooled standard error = 0.2021. The probability of $t = -1.67$ with 378 degrees of freedom is 0.095. We conclude that differences as large as seen here between tonnages of porphyry copper deposits in China and worldwide porphyry copper deposits happen by chance alone about 10 percent of the time. Therefore, we accept the global model as appropriate for undiscovered porphyry copper deposits of Yulong, China. Similar results were observed for Cu grades ($p = 0.33$).

AN EXAMPLE

In an assessment of undiscovered mineral resources of Nevada (Cox et al., 1996; Singer, 1996), it was assumed that the undiscovered deposits could be represented by certain grade-and-tonnage models, but it was critical to test the appropriateness of the models to Nevada. A reasonable test was to compare the grades and tonnages of the deposits from Nevada to the global grade-and-tonnage models.

For example, there are seven known porphyry copper deposits (Yerington, SFS, MacArthur, Bear, Ely, Ann Mason, and Copper Canyon) that are defined in the same way as deposits in the porphyry copper descriptive model (Cox, 1986a) and the grade-and-tonnage model (Singer, Mosier, and Cox, 1986). In figure 6.5, tonnages of these deposits are plotted on the general porphyry copper tonnages model. There is no clear clustering of the Nevada deposits compared to the global model. To make sure that the general model is appropriate for Nevada, a t-test of the tonnages of the deposits in Nevada compared to the tonnages of the general model should be performed with logged tonnage data:

$$t = (8.175 - 8.355)/0.2572 = -0.698,$$

with 207 degrees of freedom, the mean tonnage for Nevada deposits = 8.355, the mean tonnage for the world (excluding Nevada deposits) = 8.175, and the pooled standard error = 0.2572. The probability of $t = -0.698$ with 207 degrees of freedom is 0.486. Similar plots and tests were performed on copper grade with the same general result. Thus, we can conclude that differences as large as seen here between tonnages of porphyry copper deposits in Nevada and worldwide porphyry copper deposits can happen by chance alone about 49 percent of the time. Therefore, the global model was accepted as appropriate for undiscovered porphyry copper deposits of Nevada.

AN EXAMPLE

Two Jurassic kuroko-type massive sulfide deposits were known in a part of southern Oregon where a mineral resource assessment was being prepared (Singer et al., 1983). The two previously mined and thoroughly explored deposits (figure 6.6) were found with a t-test to be significantly lower in tonnage ($p < 0.001$) than the general kuroko grade-and-tonnage model (Singer, 1993b). Clearly, a new grade-and-tonnage model was needed here.

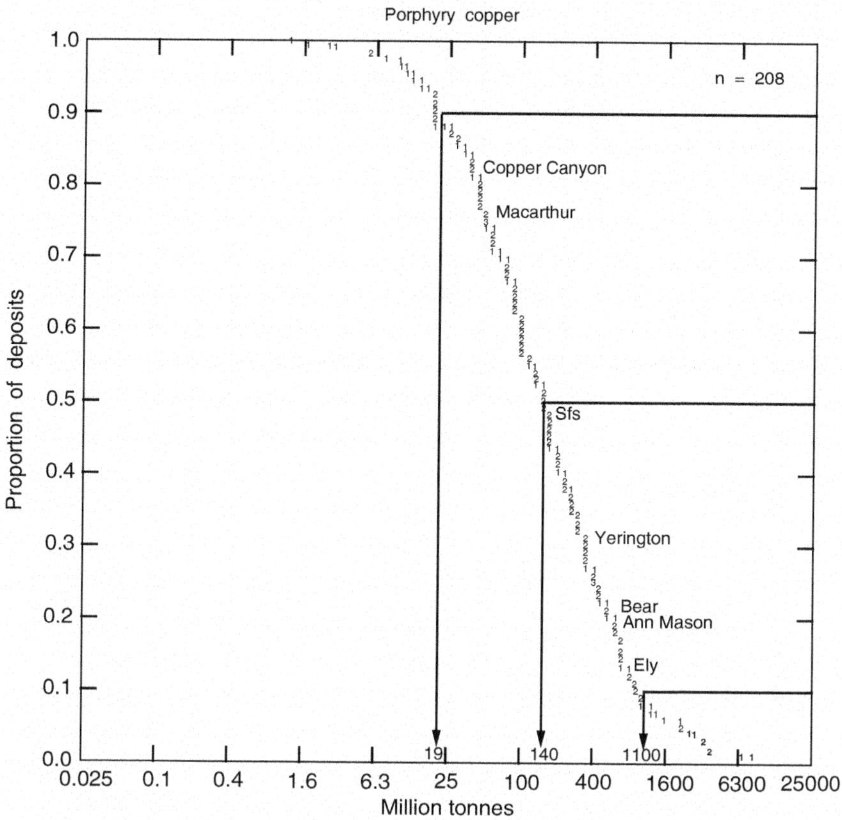

Figure 6.5 Comparison of tonnages of known porphyry copper deposits in Nevada and porphyry copper deposit model (1996).

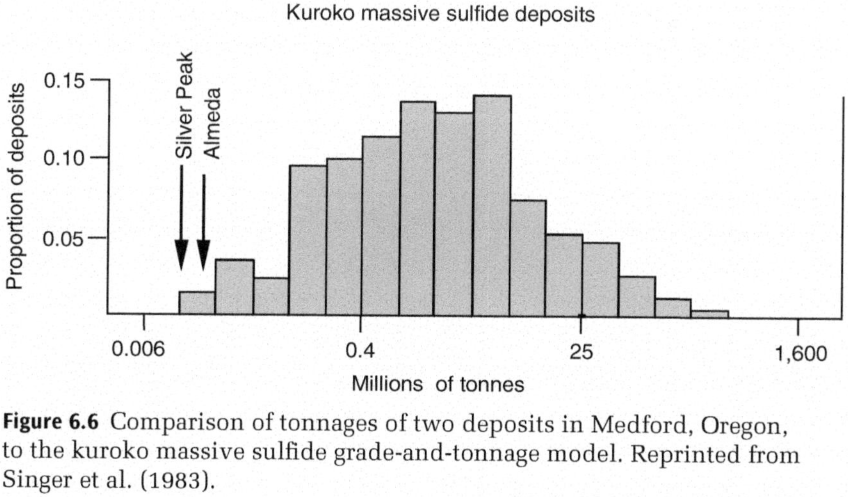

Figure 6.6 Comparison of tonnages of two deposits in Medford, Oregon, to the kuroko massive sulfide grade-and-tonnage model. Reprinted from Singer et al. (1983).

Construction of Grade-and-Tonnage Models

Construction of grade-and-tonnage models involves multiple steps, the first of which is the identification of a group of well-explored deposits that are believed by others or the modeler to belong to the mineral deposit type being modeled. A descriptive model is commonly prepared as well, and the attributes of each deposit in the group are compared with it to ensure that all are of the same type. These data include average grades of each metal or mineral commodity of possible economic interest and the associated tonnage based on the total production, reserves, and resources at the lowest possible cutoff grade. All further references to tonnage follow this definition. These data represent an estimate of the endowment of each known deposit so that the final model can represent the endowment of all undiscovered deposits.

Grade-and-tonnage Data

When planning a mine, it is common to calculate tonnage and grade at different cutoff grades. This allows engineers to plan the mine under several scenarios of cost of materials and prices for commodities. The word "reserves" applies to material that is well characterized and can be produced at a profit. Resources include reserves and additional material that is too low grade to currently be profitably produced. As prices and costs change during mining, reserves of deposit may be updated. Often, costs of mining decrease as mining takes place, and lower grade material that was not initially thought to be economic to produce will be able to be profitably mined. Grade-and-tonnage models used in three-part assessments represent the grade and total tonnage of a deposit before any material is removed. This means that current resources at the lowest cutoff grade are added to past production. Grade-and-tonnage models use resource figures, that is, grades and tonnages calculated at the lowest cutoff grade, to represent the mineralized material in a deposit in order to allow for possibly different technologies and mining costs to be assumed.

AN EXAMPLE

A porphyry copper deposit has a reported inferred plus indicated resource of 1,551 million tons with a copper grade of 1.03 percent and a gold grade of 0.35 g/t gold at a 0.6 percent copper equivalent cutoff and 3,110 million tons at 0.68 percent Cu and 0.24 g/t gold at a 0.3 percent copper equivalent cutoff grade. In this situation, the tonnage and average grades associated with the 0.3 cutoff grade would be used.

AN EXAMPLE

By the end of 1996, mining at Nanisivik in Canada had produced 104,400 short tons of lead, 1,339,000 short tons of zinc, and 15,130,000 troy ounces of silver from 14,750,000 short tons of ore. In addition, the reserves at that time were 22,300 tons of lead, 361,000 tons of zinc, and 4,810,000 ounces of silver in 4,460,000 tons of ore. The mill recovered 87 percent of the lead, 95 percent of the zinc, and 79 percent of the silver.

The premining lead grade can be calculated as

$$(104,400\,\text{st}/0.87 + 22{,}300\,\text{st})/(14{,}750{,}000\,\text{st} + 4{,}460{,}000\,\text{st})$$
$$= 142{,}300/19{,}210{,}000 = 0.74\%\ \text{lead}.$$

The premining zinc grade is

$$(1{,}339{,}200\,\text{st}/0.95 + 361{,}000\,\text{st})/19{,}210{,}000\,\text{st} = 9.2\%\ \text{zinc}.$$

The silver grade is

$$(15{,}130{,}000\,\text{oz}/0.79 + 4{,}810{,}000\,\text{oz})/19{,}210{,}000\,\text{st}$$
$$= 1.25\,\text{oz silver per short ton of ore}.$$

In metric units, the silver grade is

$$1.25\,\text{oz/st} \bullet 34.285\,\text{gst/oz t} = 42.8\,\text{g/t Ag}$$
and the deposit size is $19{,}210{,}000\,\text{st} \bullet 0.907 = 17{,}400{,}000\,\text{t}$.

Plotting the Data

The next step is to plot the data. For tonnage and most grades, a transformation to logarithms is necessary to remove skewness. Histograms, normal probability plots, cumulative frequency plots, and empirical quantile function plots are all useful, as is the examination of skewness and kurtosis statistics. Bivariate (scatter) plots of each pair of variables should also be constructed. An artificial example of tonnages and zinc grades of twenty deposits serve to demonstrate how to plot grade-and-tonnage data. Deposits in table 6.1 have tonnages reported in millions of metric tons, zinc grades in percentages, and the $\log_{10}$ of tonnages and grades are shown. A plot of average zinc grade versus tonnage (logged data) shows no problem clusters or groupings of the deposits, and the correlation coefficient ($r = 0.13$) is not significantly different than zero (figure 6.7).

In table 6.2, the deposits are sorted from lowest tonnage to highest, and a new column contains the cumulative proportion of the twenty

Table 6.1. Data on tonnages and Zn grades for example plots.

Deposit	Tons (millions)	Zn (%)	Log Tons	Log Zn
A	0.16	1.7	−0.796	0.230
B	4	2	0.602	0.301
C	4.1	5.4	0.613	0.732
D	22	5.1	1.342	0.708
E	33.4	3.22	1.524	0.508
F	9	9.5	0.954	0.978
G	5	6.4	0.699	0.806
H	0.74	10.2	−0.131	1.009
I	0.459	4.27	−0.338	0.630
J	10	6	1.000	0.778
K	10.9	8.3	1.037	0.919
L	80	9.2	1.903	0.964
M	2.68	4.9	0.428	0.690
N	3.52	5.86	0.547	0.768
O	1.39	9.6	0.143	0.982
P	29.8	6.3	1.474	0.799
Q	80	2.1	1.903	0.322
R	4.8	5.49	0.681	0.740
S	11.67	8.5	1.067	0.929
T	6.8	8.7	0.833	0.940
		Mean	0.774	0.737
		St. Dev.	0.701	0.235

deposits (20/20, 19/20, 18/20, etc.). A plot of the cumulative proportion of deposits against log tonnage represents the basic tonnage model (figure 6.8). To be consistent with other tonnage plots, major ticks on the x-axis are 0.6 units (in logarithms) and one of the ticks is at 2.0 (100 million tons). Two scales are used for the x-axis in this example—the upper scale represents the logged data as plotted and is useful for placing the 90, 50, and 10 percentiles, whereas the lower scale represents the antilog of the upper scale, which can replace the upper scale after plotting the quantiles. The 50th percentile of tonnage can be estimated by the mean of the logged data listed in table 6.1 (i.e., 0.774). The 10th quantile is estimated as the mean plus 1.282 times the standard deviation in table 6.1 (i.e., 1.672). Similarly, the 90th quantile is the mean minus 1.282 times the standard deviation (i.e., −0.124). A vertical line with an arrow at the end is placed at 0.774 on the x-axis and drawn up to the 0.5 proportion of deposit on the y-axis. A horizontal line from the vertical line is drawn to the right side of figure 6.8. Below the arrow the antilog of the mean is placed. The same procedure is used to the 90th and 10th quantiles. At this point, the logged values on the x-axis are no longer needed and can be removed. In the upper right corner of the plot, the number of deposits

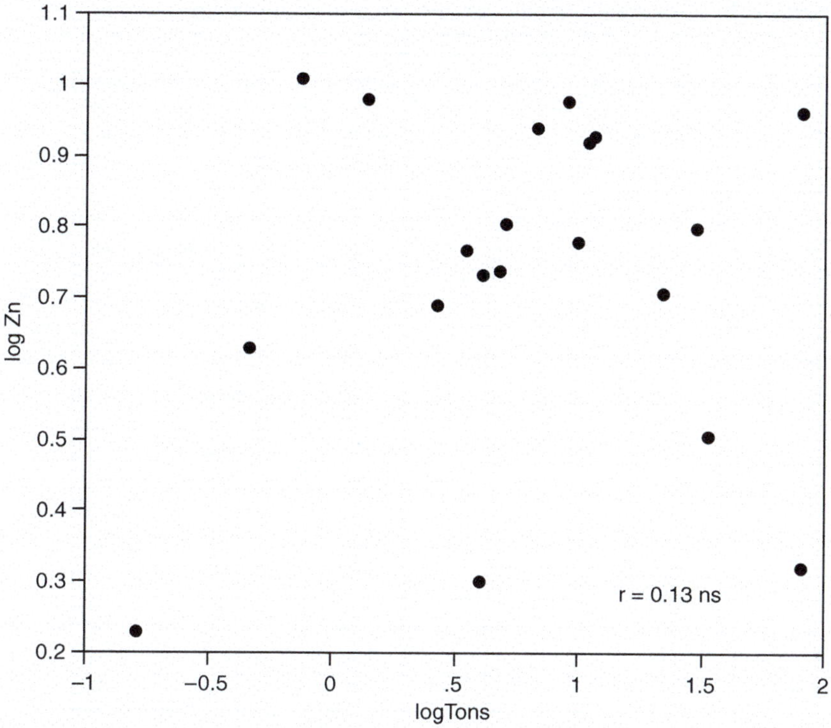

Figure 6.7 Average zinc grade ($\log_{10}$) versus deposit tonnage ($\log_{10}$) from table 6.1. The correlation coefficient (r) is not significantly different (ns) than zero.

is listed. Smoothed curves can be drawn through the three fixed quantile locations if desired.

The reader may have noticed that tonnages associated with the proportion of deposits in table 6.2 could have been used as estimates in figure 6.8. It is also possible to sort from largest to lowest tonnage and estimate the cumulative proportion of deposits—this produces estimates different than those in table 6.2. Algorithms in many computer programs provide compromise estimates. It is certainly possible to use a different algorithm to make these estimates. Because we have tested the frequency distribution of tonnage here (and in all models) and found that the distribution is not significantly different than a lognormal distribution, we have used the lognormal to estimate the quantiles. The purpose of these plots is to provide an easy way to compare deposit types and also to identify possible errors in the models.

In an example of some of the steps in constructing a new grade-and-tonnage model, two Jurassic kuroko-type massive sulfide deposits in

Table 6.2. Data sorted on tonnages with Zn grades for example plots.

Deposit	Tons (millions)	Zn (%)	Log Tons	Log Zn	Proportion
A	0.16	1.7	−0.796	0.230	1
I	0.459	4.27	−0.338	0.630	0.95
H	0.74	10.2	−0.131	1.009	0.9
O	1.39	9.6	0.143	0.982	0.85
M	2.68	4.9	0.428	0.690	0.8
N	3.52	5.86	0.547	0.768	0.75
B	4	2	0.602	0.301	0.7
C	4.1	5.4	0.613	0.732	0.65
R	4.8	5.49	0.681	0.740	0.6
G	5	6.4	0.699	0.806	0.55
T	6.8	8.7	0.833	0.940	0.5
F	9	9.5	0.954	0.978	0.45
J	10	6	1.000	0.778	0.4
K	10.9	8.3	1.037	0.919	0.35
S	11.67	8.5	1.067	0.929	0.3
D	22	5.1	1.342	0.708	0.25
P	29.8	6.3	1.474	0.799	0.2
E	33.4	3.22	1.524	0.508	0.15
L	80	9.2	1.903	0.964	0.1
Q	80	2.1	1.903	0.322	0.05

southern Oregon where a mineral resource assessment was being prepared were found to have tonnages significantly different than the general kuroko massive sulfide model (Singer et al., 1983). Examination of other Mesozoic kuroko-type massive sulfide deposits hosted in similar rocks in the western United States and Canada suggested that the tonnages and grades of the two Oregon deposits were similar to this subgroup of kuroko deposits (Singer et al., 1983). Although the frequency distribution of tonnage for this new group was determined to be significantly skewed in 1983, the model was not reexamined until 1988 during an analysis of Nevada's resources, which illustrates the iterative nature of constructing and revising deposit models. When Britannia and one other Cretaceous deposit were removed, the frequency distribution of tonnage was not significantly skewed. This new grade-and-tonnage model is associated with the descriptive model for kuroko massive sulfide deposits (Singer, 1986); however, only kuroko deposits of Triassic or Jurassic age in North America were used to construct this subset. Because many of the deposits lie in the western foothills of the Sierra Nevada in California, the name "Sierran kuroko" is given to the group (Singer, 1992). These deposits are significantly lower in tonnage than the worldwide kuroko group, but they are not different in copper grade. Although gold grades of the Sierran kuroko

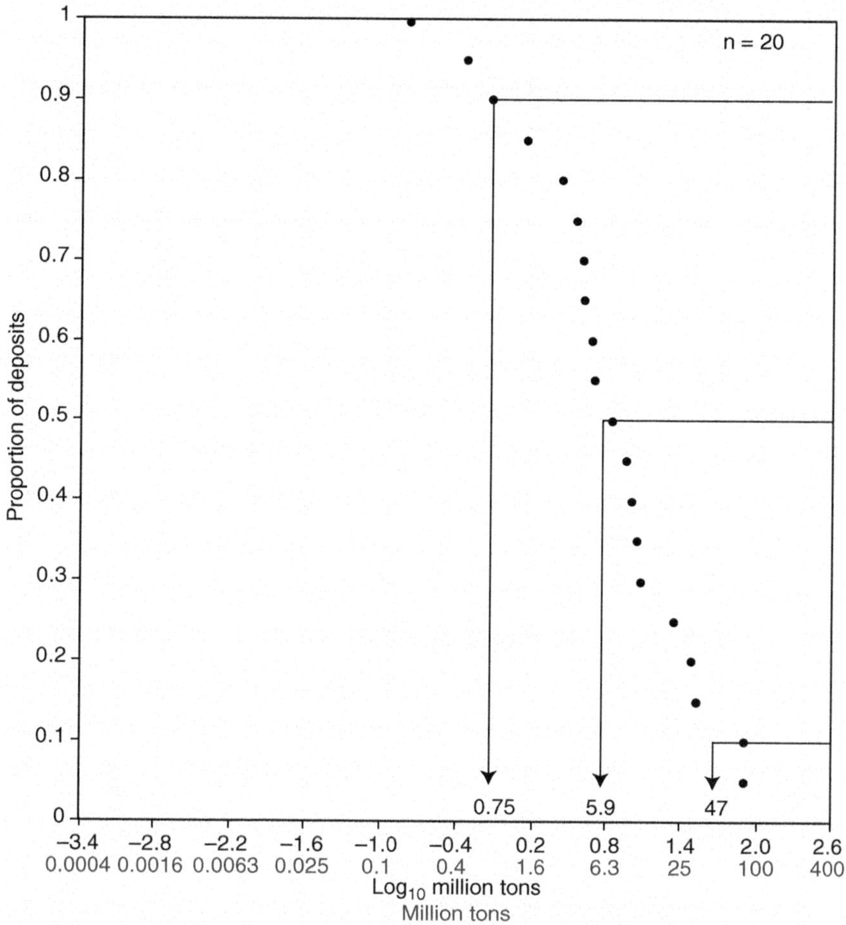

Figure 6.8 Plot of the cumulative tonnages of data in table 6.2 showing number of deposits (n) and both millions of tons and $\log_{10}$ tons.

deposits are not significantly different than the general kuroko deposits, some of the deposits in the Sierran group were primarily mined for gold in their early production stages, and much of this early production was probably unreported. The reason for the lower tonnage and possible higher gold grades is not known. Speculative reasons include (1) a short-lived volcanic arc and therefore not enough time to form large deposits, (2) poor preservation of deposits, (3) shallow emplacement of deposits, and (4) postdeposition tectonic tilting and erosion. Construction of this model demonstrates (1) situations where regional models are justified, (2) the usefulness of plotting data, and (3) the iterative nature of building grade-and-tonnage models.

The Lognormal Distribution

Use of probability paper or Q-Q plots provides the experienced analyst a quick method of judging whether the population may feasibly be lognormal, but they should not be regarded as a rigorous statistical test. For sixty-seven deposit models, Singer (1993a) rigorously tested tonnages for lognormality using the Dallal and Wilkinson (1986) modified Lilliefors test and the joint moment-ratio ($\sqrt{b_1}$ and b_2) tests shown by Shapiro, Wilk, and Chen (1968) to be quite sensitive to nonnormality. Both sets of tests are recommended by Stuart, and Ord (1991) and the joint moment-ratio ($\sqrt{b_1}$ and b_2) tests are recommended by Aitchison and Brown (1963). Singer (1993a) shows that five of the sixty-seven tonnage distributions are significantly different (either skewed or peaked) than lognormal at the 1 percent level. In addition to the above results as an empirical basis for believing that the lognormal distribution is an appropriate model for most observed mineral deposit tonnages, a long and distinguished history of scientific publications provides a theoretical and empirical basis (Aitchison and Brown, 1963; Allais, 1957; Brinck, 1967; Matheron, 1959; Rasumovsky, 1940).

Typically, a lognormal distribution fits observed distributions of homogeneous populations of variables representing weights, lengths, volumes, and grades of trace quantities. It tends to not fit distributions of elements that have grades greater than about 10 percent, such as Fe, Mn, and Al, that can sometimes be made approximately normally distributed by the square root transformation, but they commonly have mineralogical limits that reduce the usefulness of transformations. Barite ore grades are an example of distributions skewed to the left and not easily transformed to normality.

Sources of Errors in Building Models

Mineral deposit modeling errors can be caused by mixed geologic environments, poorly known geology, data recording errors, mixed deposit and district data, mixed mining methods, or incomplete production/resource estimates. Constructing grade-and-tonnage and descriptive models is an iterative process wherein the modeler attempts to identify and remove these errors. The plots and statistics help discover if the data contain multiple populations or outliers.

Based on our experience with a large number of models, deviations from lognormality, outliers, or subgroups are all cause for reexamination of the data. Also suggestive of problems are large standard deviations for tonnage, such as those greater than 1.0, and significant correlations between tonnage and grade. If any of these conditions exist, the data should be checked for correctness of data entry, data reporting errors, mixed sampling units, and, lastly, correctness of the geologic reasoning that led to the classification of the individual deposits. If subgroups of data exist, one or more geologic attributes of the subgroups probably will be different, which suggests the descriptive model may need reexamination. In most cases, the process of model building requires multiple iterations. Two related factors that make revisions of grade-and-tonnage models necessary are the use of grade-and-tonnage estimates from incompletely explored deposits and the use of data on new or incompletely understood deposit types.

Although it is not possible to guarantee that a model will never be revised, a model will probably be stable if (1) tonnage and grades of commodities that constitute less than 10 percent of the ore are not significantly different from a lognormal distribution, (2) at least twenty deposits are used, (3) standard deviations for tonnage are less than 1.0, and (4) there are no significant correlations between tonnage and grade. Only two deposits are required to construct a grade-and-tonnage model, but the statistical estimators will not be very stable—with typically skewed distributions of raw grade-and-tonnage data, twenty or more deposits are recommended for stable estimates. Figure 6.9 gives an example of the value of plots and statistics: the distribution of tonnage is significantly skewed, as demonstrated by the seven orders of magnitude range of tonnages in the plot, has a standard deviation greater than 1.0 (1.7), and is correlated with gold grade ($r = -0.7$) in a proposed gold skarn model. In this example, reasons

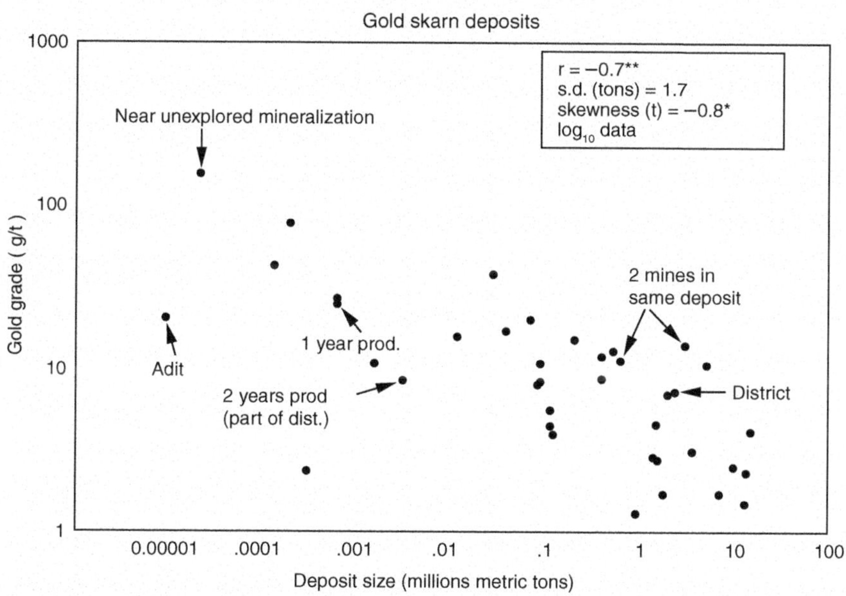

Figure 6.9 Average gold grades versus deposit tonnage for gold skarn deposits used in a grade-and-tonnage model. Data from Theodore et al. (1991). r = correlation coefficient; s.d. is standard deviation of $\log_{10}$ data. *Significant at the 5 percent level; **significant at the 1 percent level.

for these problems can be identified as incomplete records of production or resources, mixed deposit and district data, and mixed mining methods. The wide range of tonnages is linked to the range of data types from adits to districts.

For some deposit types, such as placer Au, a correlation between tonnage and gold grade is due in part to the effects of different, but largely inseparable, mining methods that have been used (Bliss, Orris, and Menzie, 1987). A small subset of placer data, for which mining method (not representative of all data) is known, shows that the correlation between tonnage and gold grade is significant if all data are included, but it is not significant if placers predominantly mined by dredging are removed from those mined by small-volume methods (figure 6.10). In such situations, the model will remain unmodified until the effects of mining method can be related to grades and tonnages and the descriptive model can be revised to reflect where different mining methods are applicable, such as low relief areas for dredges.

Sampling Unit Deposit Rules

An important consideration at the data gathering stage is the question of what the sampling unit should be (Singer, 1993b). We would like to have our sampled population (deposits) represent the target population (undiscovered deposits) about which we will be making inferences. The geologic and mining literature contains many terms such as district, zone, ore body, lens, shaft, vein, bench, and mine that might be considered as possible sampling units. These terms are applied in different ways by different groups at different points in time, making them undesirable as our sampling unit. For the most part, the data included in published grade-and-tonnage models represent individual deposits, but in some instances such data are mixed with data representing districts. Grade-and-tonnage data are available to varying degrees for districts, deposits, mines, and shafts. In many cases, old production data are available for some deposits and recent resource estimates are available for other deposits. Probably the most common error in constructing grade-and-tonnage models is mixing old production data from some deposits with resource data from other deposits (figure 6.9). It is extremely important that all data used in the model represent the same sampling unit because mixing data from deposits and districts or old production and recent resource estimates usually produces bimodal or at least nonlognormal frequencies and may introduce correlations among the variables that are artifacts of the mixed sampling units (figure 6.9). Models constructed using data from mixed sampling units are of questionable value because the frequencies of tonnage and grade observed are directly related to the proportion of deposits from each sampling unit and are unlikely to be representative of the proportion in the undiscovered deposits being estimated.

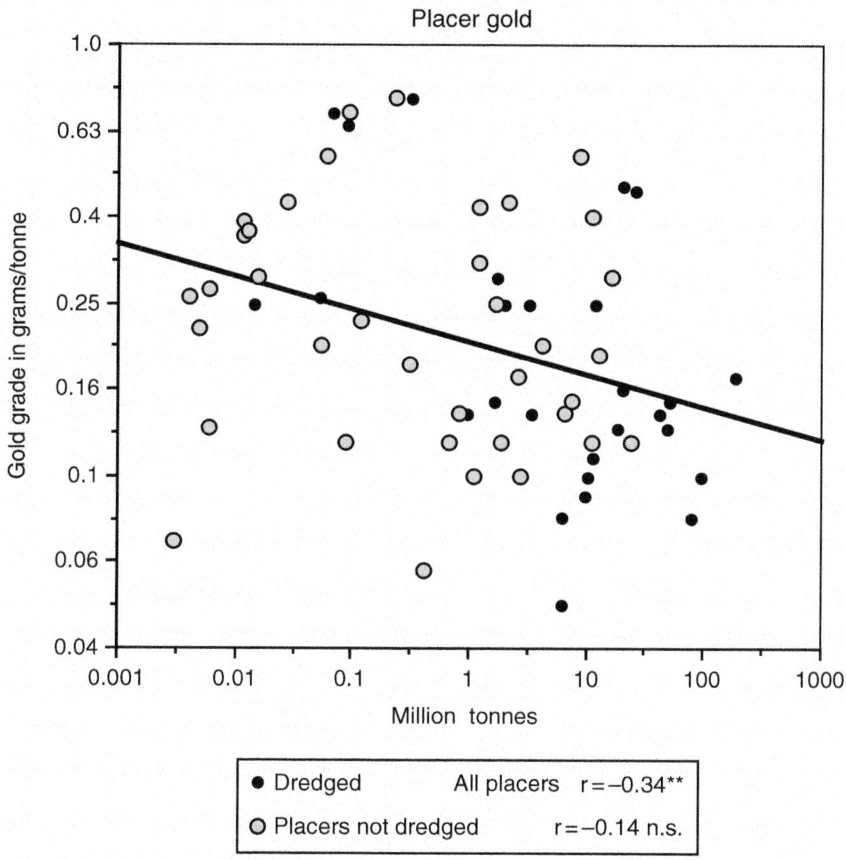

Figure 6.10 Average gold grades versus tonnage for deposits in the placer Au grade-and-tonnage model. The correlation is not significant (n.s.) for placers that have not been dredged but is significant (**) for all data including placers dredged.

For deposit models, we try to use a spatial rule to determine which ore bodies should be combined so that we can have a consistent sampling unit that can be applied to the undiscovered deposits. For example, ore bodies of both kuroko and Cyprus-type massive sulfides were combined into single deposits based on a 500-m rule of adjacency (Mosier, Singer, and Salem, 1983). Effects of map scale on the possible definitions of a deposit are demonstrated by the kuroko deposits of the western part of the Hokuroku district (Ohmoto and Takahashi, 1983) in Japan (figure 6.11). It is clear from figure 6.11 that kuroko deposits could be defined as very small bodies depending on the scale of the map. When the 500-m rule of adjacency for these deposits is applied, only three deposits are defined. All of the smaller parts were added together, as were their grades and tonnages, for the grade-and-tonnage model. In another example, deposits in a recent report on porphyry copper deposits (Singer, Berger, and Moring, 2008) consistently follow two operational rules to define a deposit in the grade-and-tonnage model: (1) all mineralized rock or alterations within 2 km were combined as one deposit, and (2) grade-and-tonnage data reported for deposits, including average grade of each metal and associated tonnage, were based on total production, reserves, and resources at the lowest possible cutoff grade for thoroughly delineated deposits. These operational rules, including the spatial rules defining the sampling unit, are necessary for defining deposits in order to ensure that deposits in grade, tonnage, and deposit-density models consistently correspond to both discovered and estimated undiscovered deposits. Some examples illustrate the effects of the application of this rule. In many compilations of mineral deposits, Chuqui Norte, Exotica, Radomiro Tomic (Pampa Norte), and Chuquicamata in Chile are reported as separate deposits, whereas in Singer, Berger, and Moring (2008) they were reported as one deposit because of the 2-km rule. El Pachon in Argentina and Los Pelambres in Chile are two parts of the same deposit, frequently reported as separate deposits. Information about such rules is available in the references. Although the specific distance for these rules is arbitrary, the distance should be such that information is available to enforce the rule consistently. Application of spatial rules can significantly affect a grade-and-tonnage model—not applying such rules leads to a poorly defined model and possibly introducing biased estimates of the grades and tonnages of the undiscovered deposits.

Additions to Reserves

Poorly reported grade-and-tonnage data used to construct grade-and-tonnage models are another possible problem. Mining enterprises must spend money and time to prove reserves and resources. As an economic activity, proving reserves and resources should be conducted to determine if the deposit is economic to mine at an acceptable level of risk and to plan the mining activity. It would be unwise to expend money now when there is no benefit until the distant future. Also, in some localities, reserves are taxed, so there is an incentive to not drill and report reserves until necessary. Thus, for most kinds of mining and in most localities, reserves and resources are drilled and reported over a period of years as mining progresses. Because the tonnages used in grade-and-tonnage models represent the total of production, reserves, and the various categories of indicated and inferred resources, the conversion of some of the inferred or indicated into the reserve category should not significantly change the tonnages in the model because material is just moving from one to another category within the total tonnage. Uncertainties about the values of the various categories might be reduced by additional drilling. However, if there is a large increase in the price of a

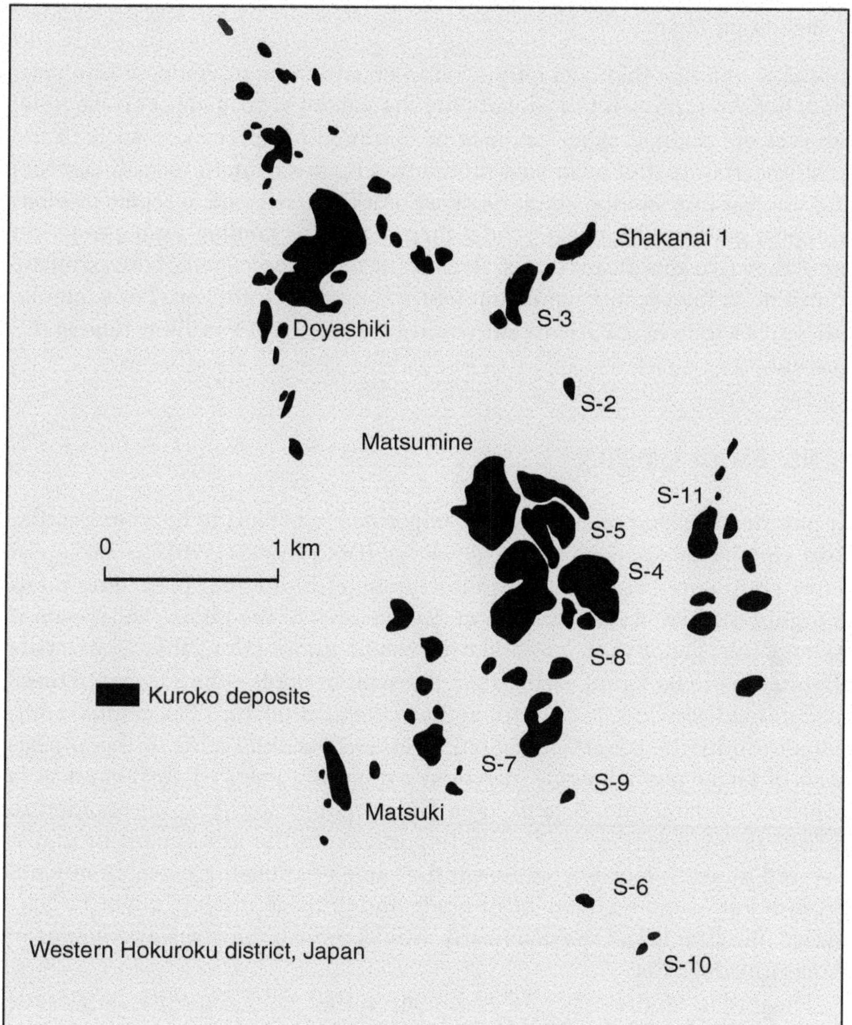

Figure 6.11 Kuroko massive sulfide deposits of the western part of the Hokuroku district. After Ohmoto and Takahashi (1983).

commodity, it is possible that new resources might be identified based on a lower cutoff grade than was ever considered before, thus causing an addition to resources. There are also deposit types, such as sediment-hosted gold, that are initially so poorly understood that parts of the deposit, such as deep sulfide ore, are not drilled for many years (Singer, 1993b). Not recognizing these situations will lead to incomplete and therefore biased grade-and-tonnage models.

Significant Digits

It is easy to forget that data employed in construction of grade-and-tonnage models have quite a bit of uncertainty associated with them. For example, reserves are defined as an estimate of unmined resources known with the least uncertainty. But even past production figures contain recording errors and unclear information about recovery. For these reasons, it seems prudent to report grade-and-tonnage model figures such as median tonnages to not more than two significant digits. Depending on the precision of the estimate, rounding to the second significant figure should be sufficient. For example, 10,743,724 tons at 7.23 per cent could be stated as 11 million tons at 7.2 percent.

Size-Biased Sampling

In petroleum resource exploration, larger deposits tend to be found earlier than smaller deposits as a play develops (Drew, 1990). With exception of a few petroleum-bearing basins, the volume of individual petroleum pools is highly correlated with surface projection area of the pools. Thus, even if geology and geophysics provided no useful information, the larger pools should tend to be found earlier than the smaller pools—the success of these disciplines (Drew, 1990) only strengthens the relationship. The gradual additions to individual petroleum pool reserves over time adds to the appearance of larger pools being discovered earlier, but even when this effect is adjusted for, larger pools tend to be discovered earlier. This observation in petroleum exploration is of some importance to the assessment of undiscovered mineral deposits because if the same relationship exists in mineral exploration, some adjustment of grade-and-tonnage models might be necessary. Finding larger deposits early would reduce the sizes and values of remaining deposits.

The order of discovery of sediment-hosted gold deposits in Nevada demonstrates how an additions-to-reserves situation can masquerade as size-biased sampling (figure 6.12). By 1987, deposit size was becoming significantly smaller as exploration progressed. The distribution of tonnage was significantly skewed toward larger tonnages due to the two very large deposits that were found relatively early in the exploration process (figure 6.13). No geologic reason has been found to distinguish these large deposits from the other deposits, but these two deposits appear to be more thoroughly explored both laterally and vertically than most of the other deposits plotted. The deep sulfide mineralization first recognized in these two deposits was initially considered as refractory and therefore uneconomic. This suggests that many of the other deposits will eventually be found to be much larger than now estimated. The problem in constructing a grade-and-tonnage model of undiscovered sediment-hosted gold deposits is to somehow

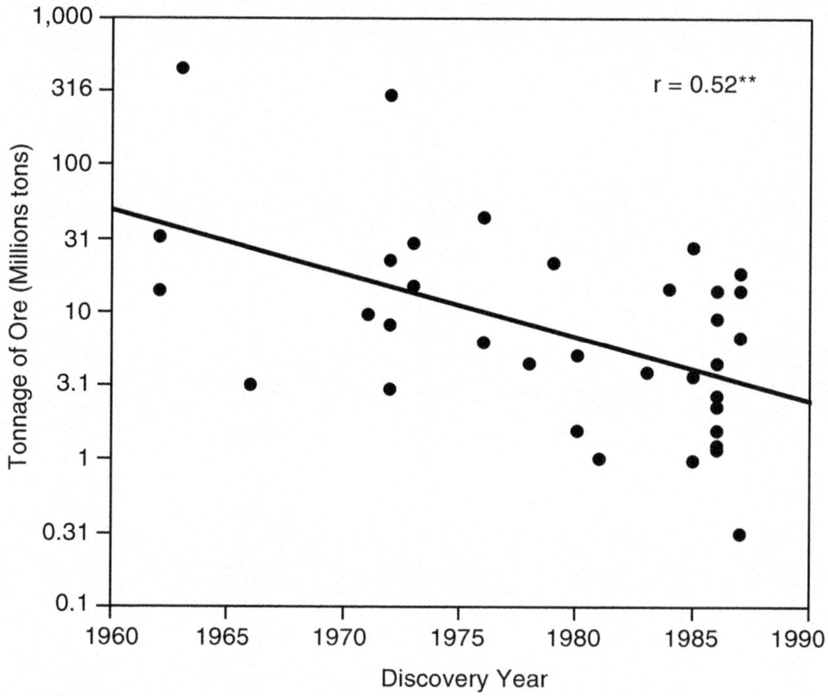

Figure 6.12 Discovery order versus deposit size of sediment-hosted Au deposits in Nevada to 1987. **Significant at the 1 percent level.

represent the deposits after the additions to reserves while only having data from two completely drilled deposits. One solution is suggested in figure 6.13, where the tonnage curve is moved over to cover the two largest deposits and yet retain the same slope curve. This solution is far from ideal because two deposits are controlling most of the model, but using the unadjusted model would clearly misrepresent the tonnages of undiscovered deposits even more poorly.

Most economic geologists believe that there should be an inverse relationship between mineral deposit discovery order and deposit size, but few studies document this relationship for mineral deposits (Stanley, 1992). One study on metal-bearing deposits showing a pattern of finding larger deposits early in the exploration process used mercury deposits in California (Chung, Singer, and Menzie, 1992). Epithermal gold deposits in Nevada and carbonatite deposits in Brazil show no relationship between size and discovery order—in both cases, however, large discoveries were made late in the exploration process in areas of difficult access. In the case of carbonatite deposits in Brazil, perhaps the largest deposit, Seis Lagos, was discovered in recent years in the remote headwaters of the Amazon River. The larger Nevada

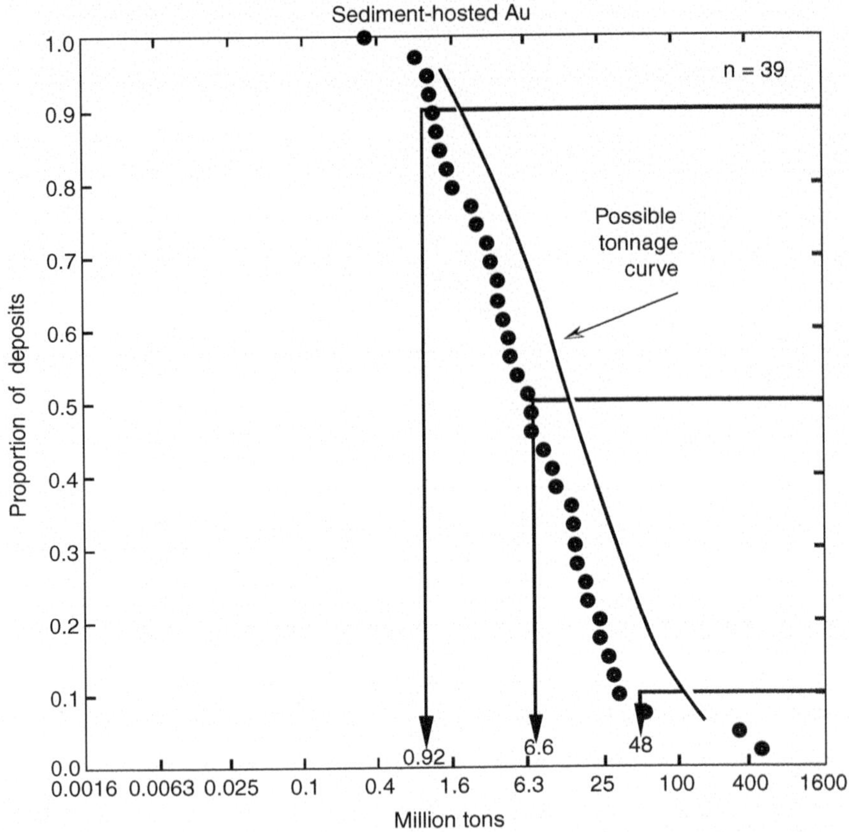

Figure 6.13 Possible adjustment of tonnage model for size-biased sampling. If the tonnage model is lognormal with the same standard deviation and the two largest deposits are on the curve, then the curve must move significantly to the right.

epithermal gold deposits discovered in 1890–1910, such as Tonopah and Goldfield, are located off the commonly used migration paths to California during the California gold rush, which is where most Nevada epithermal deposits discovered in 1840–1870 are located.

For some deposit types, such as porphyry copper, there is evidence that the larger deposits should be discovered earlier than smaller deposits (Singer and Mosier, 1981b). For both covered and exposed porphyry copper deposits, the area of the sulfide zone (disseminated pyrite) is an important determinant of discovery chances. Because there is a strong positive relationship between area of sulfides and the copper contained in the deposits, larger deposits should be discovered earlier than smaller deposits (Singer and Mosier, 1981b). However, this is only true within fixed exploration settings such as an exposed permissive rock that has all parts equally accessible. Where the exploration setting changes, for example, looking under shallow cover with a particular technique, then the process of finding larger deposits starts over—Boldy (1977) demonstrates the effect of exploration method on deposit size discovery order in the search for massive sulfide deposits.

A notable difference between the petroleum and mineral industries is the amount of information available on exploration that might be used to demonstrate size-biased sampling. Among the consequences of the paucity of data on exploration history on minerals exploration in some parts of the world is the difficulty of making convincing cases for size-biased sampling and, perhaps more important, the difficulty in constructing unbiased grade-and-tonnage models that take this possible bias into account. Where data permit, one possible approach would be to construct grade-and-tonnage models that are conditioned (with linear regression) on discovery order.

Economic Effects on Grade-and-Tonnage Models

Deposits strongly suspected to be small or very low grade are seldom sampled well enough to be characterized in terms of grade and tonnage; therefore, one would expect that the sample of many deposit types would be truncated by economics. Effects of economic filtering should be most evident in plots of grade versus tonnage for which the combination of low grade and low tonnage should be missing. For almost any conceivable distribution of grades and tonnages before economic filtering, the removal of low-grade and low-tonnage deposits due to economics would cause a negative correlation in the remaining data because the lower grade and lower tonnage deposits would not be reported (Harris, 1984). The uncommonness of negative correlations in more than sixty-five published grade-and-tonnage models (Singer, 1993a) suggests economic filtering is not severe. At least 40 percent of the deposits used in these models are noneconomic today. For example, at least 50 percent of the deposits used in the grade-and-tonnage models for porphyry copper deposits have never been developed, even though most were explored more than fifteen years ago (figure 6.14). The expected missing low-grade and low-tonnage deposits are present, and the low positive correlation is significant at the 5 percent level—hardly evidence of economically filtered data. Thirty of the thirty-three porphyry Mo, low-F deposits in that grade-and-tonnage model (90 percent) have never been developed. The majority of the 435 podiform chromite deposits from California and Oregon were mined only when there was a subsidy. Perusal of the figures in Cox and Singer (1986) and Bliss (1992a) reveals many examples of both small deposits and low-grade deposits. There is also the question of whether possibly missing small deposits would have any effect on an assessment.

Potential metal supply is dominated by the very few largest tonnage deposits, as shown by Singer and DeYoung (1980), who also pointed out that

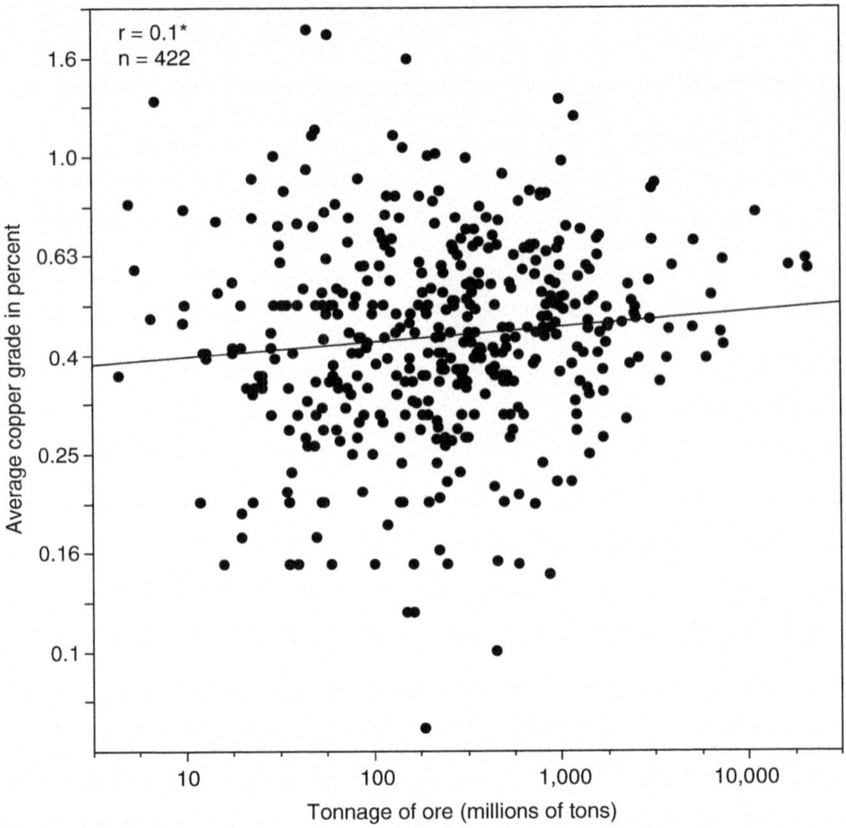

Figure 6.14 Average copper grade plotted against tonnage in all porphyry copper deposits. The correlation coefficient (*r*) is positive and significant (*) at the 5 percent level for the 422 (n) deposits. Reprinted from Singer, Berger, and Moring (2008).

inverse correlations between grade and tonnage are surprisingly rare. Therefore, a low-grade deposit will not necessarily be large. As a consequence, most low-grade deposits are not likely to have huge resources, and omitting a few low-grade or small-tonnage deposits will not seriously degrade the predictions of potential supplies of most commodities.

From the preceding discussion, it is clear that most of the published grade-and-tonnage models include a significant proportion of noneconomic deposits and that, in most cases, low-grade or low-tonnage deposits not included in the models would have negligible effect on potential supply estimates. In the view of most economic geologists, however, low-grade and particularly low-tonnage deposits are underrepresented in the models. The missing low-grade and small-tonnage deposits suggest that grade-and-tonnage models represent a biased sample of the large number of low-grade or small-tonnage occurrences and prospects found during exploration. This difference between the population represented by the grade-and-tonnage model and the population that may exist in the earth must be considered when the number of undiscovered deposits is estimated (see chapter 8).

It might be argued that the grade-and-tonnage models should be extended to include not only deposits but also occurrences that are typically very small concentrations of a mineral. If the problem of possible biases due to incomplete exploration of these occurrences is neglected, then it is possible to construct such models; the tonnage model would of course have a much lower median. Because quantitative assessments require that the estimated number of undiscovered deposits be consistent with the grade-and-tonnage model, the process of estimating the number of deposits might be more difficult because of the much larger number of "deposits" (including occurrences) to be estimated. An economic analysis of the results of this assessment would show that the occurrences and probably some of the estimated undiscovered deposits would be uneconomic. Thus, the effect of including occurrences in the grade-and-tonnage model would be to increase work in the assessment but yet not affect the final answer in any way.

Intradeposit Grade Variation

Data used to construct the grade-and-tonnage models have differing and frequently unknown cutoff grades. In principle, improved price or productivity should lower the cutoff grade, which in turn should add new deposits that have average grades above the new cutoff and increase the tonnage of those deposits currently under production. Lasky's (1950) analysis of the relationship between cumulative tonnage and average grade of mineralized material demonstrated that different cutoff grades can significantly change total tonnages and average grades. The close correspondence of Lasky's equation to observed data over the range of grades for which data exist has been shown to be a consequence of the lognormal distributions of grades (Matheron, 1959). However, as shown by DeYoung (1981), projection of Lasky's method to lower grades is limited because the mathematical formulation predicts physically impossible situations below some limiting grade.

Taylor (1985) combined the theoretical aspects of the lognormal distribution with actual examples and economic analysis to show how cutoff grades can, in practice, affect grades and tonnages. He concluded that the cutoff grade must be near the median of the deposit population in order to recover a reasonable proportion of the metal content in a tonnage fraction that is sufficiently large to have spatial continuity and to be minable. He also observed that many cutoff grades of mines are located at or near the deposit population medians. Thus, although wide variability in tonnages and average grades may result from changes in cutoff grades, in practice operators are limited by economics and by the consequences of needing spatial continuity and with the lognormal distribution to a rather narrow range of cutoff grades. Exceptions may exist, however, owing to differences in mining methods that significantly affect operating costs such as the very low costs of dredge mining and heap leaching for gold. Although further work is needed to define the relation of cutoff grade to these models, the effect of cutoff grades on grade-and-tonnage models may not be as pronounced as suspected provided the mining method is the same.

7

Delineation of Permissive Tracts

Perspective

The nonuniform global distributions of metals discussed in chapter 2 are also evident within most countries. Knowledge of the spatial distributions of mineral resources is invaluable in planning. In order to be able to consistently assess the undiscovered mineral resources of regions, as the second part of three-part assessments, areas should be delineated where geology permits the existence of deposits of one or more specified types. These areas, called permissive tracts, are based on geologic criteria derived from deposit models that are themselves based on studies of known deposits outside and perhaps within the study area. Thus, deposit models play the central role in identifying relevant information and in integrating the various kinds of information to delineate permissive tracts. Permissive boundaries are defined such that the probability of deposits of the type delineated occurring outside the boundary are negligible, that is, less than 1 in 100,000.

Areas are excluded from these tracts only on the basis of geology, knowledge about unsuccessful exploration, or the presence of barren overburden exceeding some predetermined thickness. A geologic map is the primary local source of information for delineating tracts and identifying which are permissive for different deposit types. Map scales affect the quality and nature of information available for delineations and determine the extent to which geologic units are combined and how cover is represented. Probably the second most important kind of information is an inventory of known deposits and prospects in and near the region being assessed. Tracts may or may not contain known deposits. Because of incomplete deposit descriptions, it often is difficult to identify deposit types for many prospects, occurrences, and some deposits, but those that can be identified increase confidence in domains delineated for the deposit type. Typed prospects may indicate the possibility of some deposit types where the type had not been expected or place limits on the kinds and sizes of deposits that could occur elsewhere. The map of deposits and occurrences classified into deposit types then serves as a check on the accuracy of the delineation of tracts permissive for types rather than a determinant of the delineation. Geochemistry of stream sediments, rocks, or soils may suggest deposit types and aid delineation of domains for some deposit types. Geophysical tools contribute by identifying extensions of permissive rock units under cover and identifying rock units in poorly mapped areas; in some cases, geophysics can identify favorable rock units, such as hydrothermally altered rocks. Permissive boundaries are reduced only where it can be firmly demonstrated that a deposit type could not exist.

Early in an assessment it is important to review available deposit models for deposit types that *may* be present in the region. The focus should be on what types of deposits *could* be in the region rather than upon what types of deposits are *known* to occur.

Permissive versus Favorable

It is natural to want to order the tracts into those that are more favorable and tracts that are less favorable. Delineation of favorable areas frequently is applied in different ways by different people because of the difficulty of defining commonly acceptable operational rules for the term "favorable." Favorableness of different tracts can be captured by the estimated number of undiscovered deposits of various types and by the possible value of those types. In three-part assessments, it is desirable to subdivide a permissive tract into two or more new tracts that have different kinds of information, different numbers of undiscovered deposits, or possibly different amounts of uncertainty about the number of deposits. One delineation strategy is to move boundaries outward from known deposits (figure 7.1). This might be considered the delineation of favorable areas in that known deposits are sometimes viewed as more likely to be near undiscovered deposits. However, in three-part assessments we try to delineate permissive areas, which can be considered as close to a yes/no decision as is practicable. Although favorable areas (in the sense of a larger number of undiscovered deposits) are a subset of permissive areas, they represent very different concepts because they represent a range of degrees. Their boundaries will coincide only if exploration coverage is very thorough and completely effective—a fairly unusual situation. Designation of a tract as permissive does not imply any

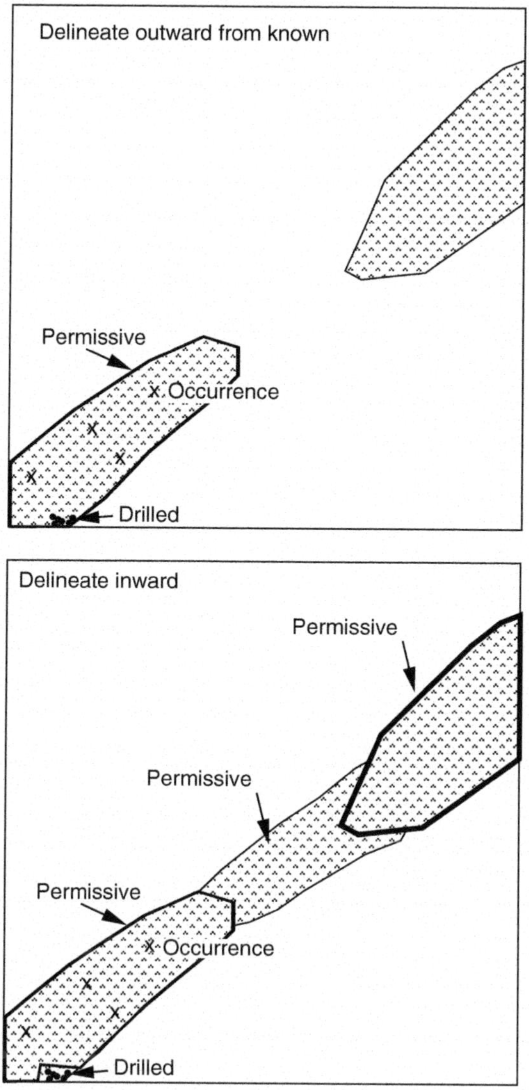

Figure 7.1 Delineation of permissive tracts. The upper plot shows the strategy of moving outward from known deposits; the lower plot shows the strategy of moving inward from permissive rocks. Moving inward is more likely to show all permissive tracts, including those without known deposits and covered settings.

special favorability for the occurrence of a deposit, nor does it address the likelihood that a deposit will be discovered there if it exists. Favorableness for a deposit type is represented by the number of undiscovered deposits that are perceived to exist in a tract.

Three strategies have been used to develop a list of candidate deposit types in the early stages of delineation of permissive tracts: (1) comparison with areas of broadly similar geology, (2) projection of deposit types from adjacent regions whose geology extends into the region, and (3) use of deposits associations to infer additional types that have yet to be discovered from the presence of associated types. The boundaries of the regions must be consistently defined, and the tract boundaries need to have been defined consistently with the region boundaries. A region can be delineated as permissive for a deposit type without having any known deposits of the type (figure 7.1). The various kinds of information available for an assessment play different roles in the delineation process.

Types of Information

Geology

A geologic map is the primary source of information for delineating tracts and identifying which tracts are permissive for different deposit types. Descriptive deposit models identify the tectonic setting and geologic environment of each deposit type. Understanding and recognizing the geologic settings associated with different deposit types require an experienced geologist and a quality geologic map. The best available map that is consistent with the scale of the final assessment map typically becomes the base map for the assessment. Identification of the "best" map is a function of whether a map differentiates rock and age units that have a bearing on recognizing the geologic setting of a deposit type. Geologic cross sections are essential to projecting surface geologic relations into the subsurface. Also useful are maps that identify the extent of cover. Although the most recent geologic map typically is the most desirable, this is not always true.

ASSESSMENT EXAMPLE

Working together, the geologic surveys of Argentina, Chile, Colombia, Peru, and the United States used a three-part assessment to delineate permissive tracts and to make probabilistic estimates of the amounts of copper, molybdenum, silver, and gold in undiscovered porphyry copper deposits in the Andes (Cunningham et al., 2007, 2008). The assessment represents one part of a project coordinated by the U.S. Geological Survey to assess the copper resources of the world in order to inform exploration managers, land-use planners, and policymakers. In addition to interest in the assessment that was expressed by mining companies from many countries and by government agencies in the assessed countries, interest was expressed by international development agencies because the assessment demonstrated the central role of quality geologic maps in predicting and guiding regional development.

Frequently, it is necessary to augment published geologic maps because the delineation must include undiscovered resources under cover. In these situations, the geology under cover needs to be estimated. Here geophysical information must play a central role. In some situations, known mineral deposits can help extrapolate the geology under cover.

Known Deposits

Classification of known deposits and occurrences into deposit types serves at least two purposes in quantitative assessments. A map of the typed deposits and occurrences demonstrates at least part of the spatial extent of processes that make specific deposit types (chapter 3). Locally, classified occurrences can indicate the presence of geology that is not expressed on the geologic map—skarn mineralization in several places of Nevada was used this way to identify where igneous intrusions not shown on the geologic map existed and therefore other deposit types related to igneous intrusions could occur (Cox et al., 1996). The presence of one deposit type, such as polymetallic veins, can be used to infer possibilities of other, related deposit types, such as porphyry Cu or Mo (Drew and Menzie, 1993). As a second purpose of classifying deposits, some of the many occurrences might, upon through exploration, be found to be a mineral deposit in the sense of being from the population of deposits on the model's grade-and-tonnage distributions. Mineral deposits are not the same as occurrences. Although the presence of typed occurrences indicates that the same processes that made a deposit took place, the vast majority of occurrences are, and will remain, just small manifestations.

Many large regions have thousands of records on occurrences that are difficult and time-consuming to classify because of the sparse information that has been gathered. Perhaps more important, even when properly classified, most mineral occurrence information is redundant. A small number of know mineral deposits, when carefully studied, often provides much more information than occurrence data. Many mineral resource assessments have an unstated assumption that being able to predict locations of occurrences has the same value to governments or companies as being able to predict the numbers and locations of mineral deposits. Just because many mineral deposits were, at one time, considered occurrences does not mean that knowing locations of all occurrences is a prerequisite to assessing or finding economic deposits. It is a common misconception among beginning assessors that one must have all occurrence data and that it must be classified into deposit types. It should also be remembered that some areas requiring assessments contain no known deposits or occurrences and yet may contain undiscovered deposits.

AN EXAMPLE

Estimating Undiscovered Deposits: Low-Sulfide Au Deposits in Venezuela

The Guayana Shield contains an Archean granulite metamorphic terrane and a Lower Proterozoic granite-greenstone terrane (Schruben et al., 1997). A number of deposits of Au in veins were known to occur on the shield. The first step in assessing the potential of the region to contain additional undiscovered Au deposits was to determine if the Venezuelan Au veins were similar to known deposit models. The Au occurred in veins and vein systems in greenschist-facies metamorphic rocks of greenstone belts and eugeosynclinal sequences, which is similar to the environment described in the low-sulfide Au-quartz vein (LGV) deposit model (Berger, 1986). To test the applicability of the LGV deposit model, the grades and tonnages of the twenty-seven known Venezuelan Au deposits were compared with the LGV grade-and-tonnage model and were found to be very similar. Thus the LGV deposit model seems to be a good representation of the Venezuelan Au deposits. The geoscience data available was largely at regional scale and was not sufficiently detailed to identify individual Au veins. The tract was composed of the greenstone belt and eugeosynclinal rock unit (figure 7.2). The tract has an area of 13,000 km^2.

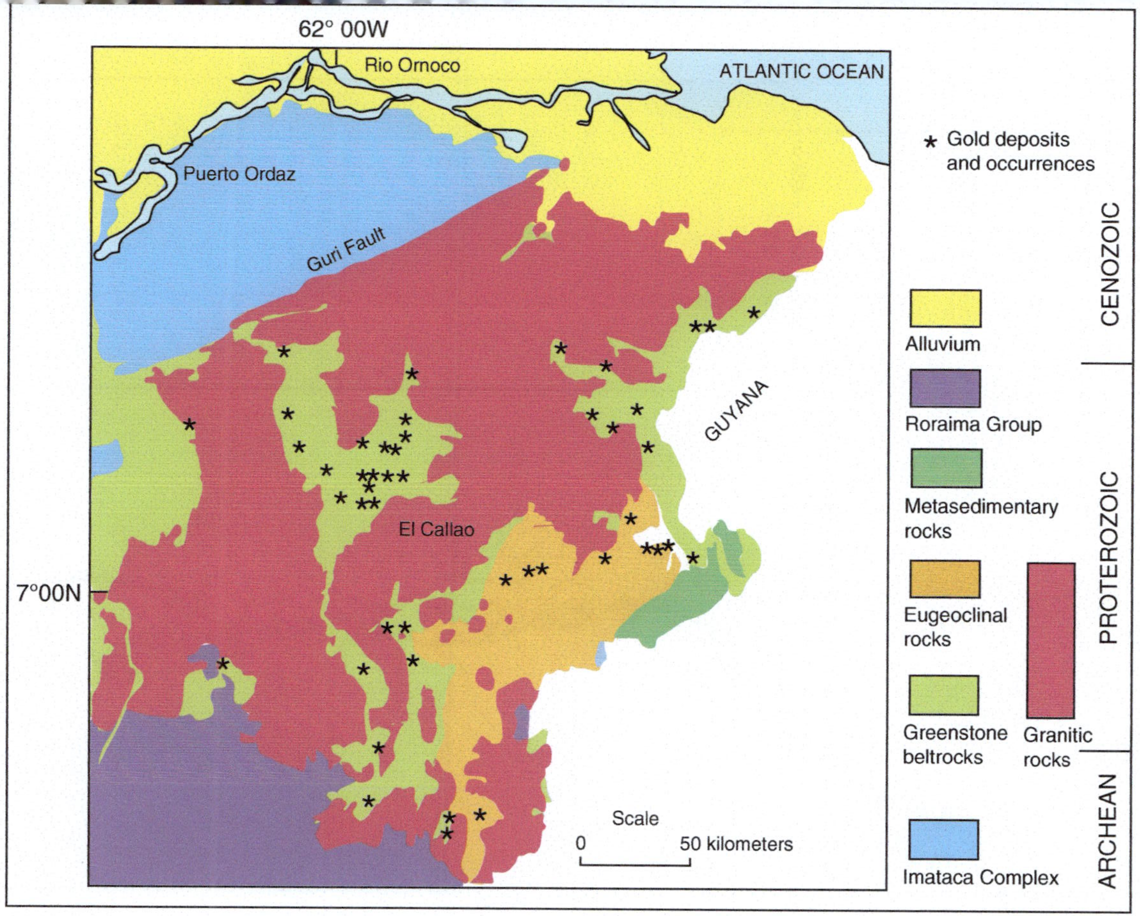

Figure 7.2 Geologic map of Guayana Shield in Venezuela with know gold deposits and occurrences. After Schruben et al. (1997).

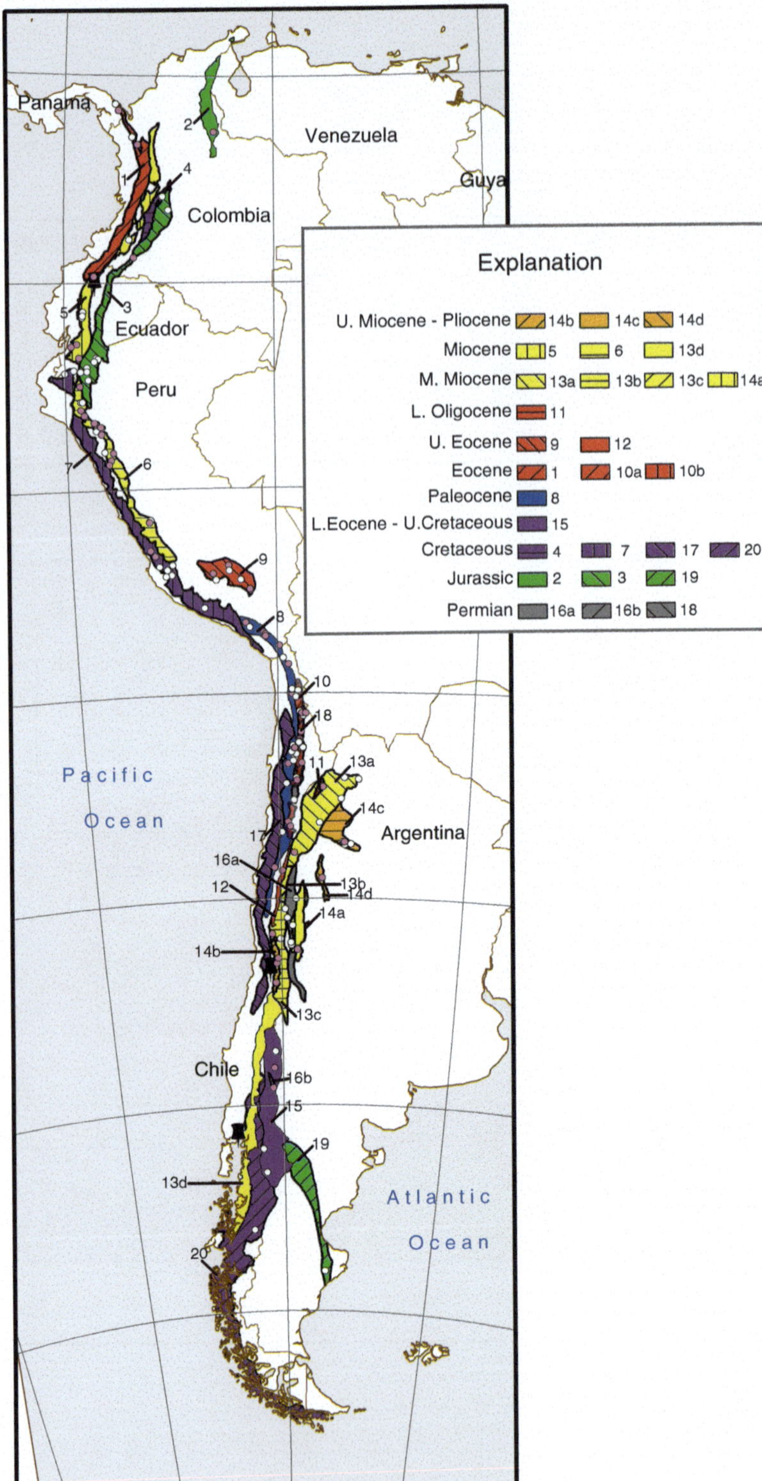

Figure 7.3 Delineation of tracts permissive for porphyry copper deposits in South America. From Cunningham et al. (2008).

AN EXAMPLE

Tracts delineated as part of an assessment of porphyry copper deposits in South America of Phanerozoic Age to a depth of 1 km below the earth's surface were confined to broad igneous arcs that formed at approximately the same time (Cunningham et al., 2008). Porphyry copper deposits form in island and continental volcanic-arc subduction-boundary zones. The twenty-six delineated tracts (figure 7.3) are believed to contain most of South America's undiscovered resources of copper. The sixty-nine known porphyry copper deposits in the Andes that meet the well-explored and 2-km spatial rules set forth (Singer, Berger, and Moring, 2005) to define a porphyry copper deposit in contrast to a prospect or occurrence are distributed in eighteen of the tracts, whereas the other eight tracts contain no known deposits. Tracts were subdivided where reasons existed to suspect spatial differences in the density or probability of occurrence of undiscovered deposits within the tract. The assessment team used the distribution of discovered deposits, appropriate prospects, similar-aged intrusive and volcanic rocks of comparable magmatic arcs, similar-aged altered rocks, fault and tectonic control, available geophysics and geochemistry, regional and deposit-model experience, and projected geology under cover. After initial data collation, the assessment of the Andes began with preliminary tract delineation, analysis of known deposits and appropriate grade-and-tonnage models, and estimation of numbers of undiscovered deposits. Although the boundaries of most of the preliminary tracts changed as the assessment progressed, having these tracts available as a starting place, together with appropriately tested, selected, or constructed grade-and-tonnage models, and preliminary estimates of numbers of undiscovered deposits using deposit-density models, greatly increased the efficiency of the international assessment team. Digital tract maps could not be constructed directly from the digital geologic maps. Geologic maps show rocks exposed at the earth's surface, whereas permissive terranes extend beneath surficial deposits of younger rocks and sediments. Maps used for delineation were at a 1:1,000,000 scale. This assessment demonstrated the value of doing a preassessment before the general meeting and of selecting qualified experts in the economic geology of the region and in assessment methods.

Geophysics

Geophysics can play several roles in delineation of permissive tracts. If the assessment is of a small area so that the resolution of the geophysical method would not miss a deposit or alteration zone that would always be recognized, then geophysics could be used directly for delineation. This is not the typical situation, however; more commonly, the scale of the survey is not able to detect deposit-size or deposit-related attributes. In a more common role, preliminary tract boundaries can be extended using geophysical surveys, such as aeromagnetics, to identify where the permissive rocks are concealed by younger rocks or sediments (figure 7.4). A geologic map is the primary source of information for delineating tracts. This presents a challenge to assessors because, for most areas, geologic maps of the rocks beneath cover are not available. Even where rocks are exposed, geophysics is commonly used to determine the extent both laterally and vertically of geologic units.

For example, analysis of regional gravity data was used by Jachens, Moring, and Schruben (1996) to estimate the thickness of Cenozoic cover in Nevada and to produce a gravity map from which the effects of thick

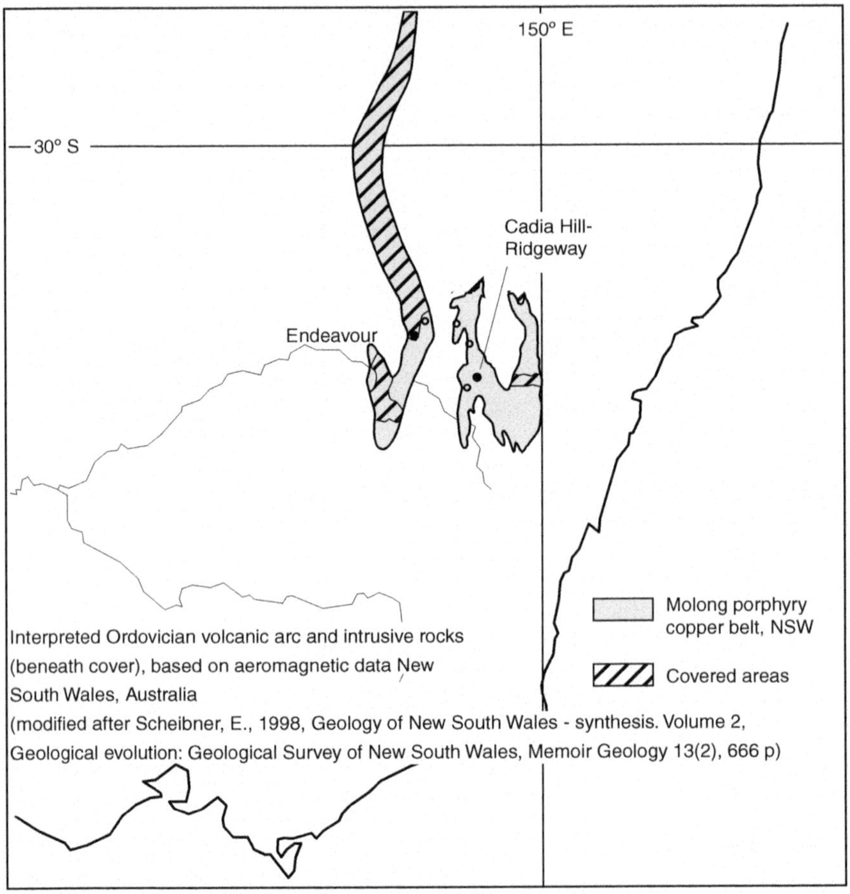

Figure 7.4 Example of delineating permissive tract under cover: interpreted Ordovician volcanic arc and intrusive rocks (under cover), based on aeromagnetic data, New South Wales, Australia. Solid circles are well-explored porphyry copper deposits, and open circles are not completely explored deposits. Modified after Scheibner (1998).

deposits of young rock and unconsolidated sediments have been removed. Their map was also used to help identify the lithologies of the concealed basement, to delineate major crustal structures and boundaries, and to identify plutons and concealed calderas, all of which can reflect geologic environments permissive for certain types of mineral deposits. In the same study of Nevada's resources, analysis of magnetic data (Blakely, Schruben, and Moring, 1996) focused on the distribution of near-surface magnetic sources in order to delineate bodies of shallowly buried magnetic rock. Typically these are Tertiary and Quaternary volcanic rocks. The information provided by this analysis affected the mineral resource analysis in that certain types of mineral deposits are associated with magnetic rocks.

Geochemistry

In most situations, geochemistry does not affect delineation boundaries. It can reduce the aerial extent of tracts only where the absence of an anomaly is definitive in ruling out the possibility of existence of the kind of deposit being assessed. Suites of elements are not unique to a single deposit type, but rather, could occur in several different deposit types. This emphasizes the point that the assessment is deposit-type specific. In uncommon situations, geochemistry can aid in delineation. Some examples might include identifying areas permissive for diamonds in regions of glacial till, or other deposit types in areas of deep weathering where the underlying rocks cannot be recognized. Geochemistry in both situations is being used to help map the permissive geologic settings.

Exploration History

Preliminary tracts should be examined to determine if parts of the tract have been previously explored for the deposits type under consideration. If parts of the tract have been explored so thoroughly that they can be confidently said to lack deposits, they should be eliminated from the tract. For some deposit types, extensive exploration might provide such evidence, but for many deposit types, only close-spaced drilling or overburden thicker than the delineation limit can be used to exclude areas. Exploration is frequently done with a specific guiding idea or model and does not necessarily test for the possibility of other kinds of deposits. Typically, exploration history plays a larger role in estimation of number of undiscovered deposits than it does in delineation (chapter 8).

Map Scale

Delineations are fundamentally based on some form of geologic map and are linked to the descriptive models. The contents of the descriptive model (chapter 3) might be correct but are not useful at the scale of the base maps used in many mineral resource assessments. In general, scales of observations in the "Geological Environment" section of descriptive models have not been identified, and yet scales are important in the application of these models in assessments. The scale of the base map also limits the information available to perform delineations and to portray delineations. Data resolution on a paper map is tied to the map's scale. Resolution is the degree to which entities can be discriminated on the map. Resolution limits the minimum size of feature that can be represented. Typically, a line cannot be drawn much narrower than about 0.5 mm. Therefore, on a 1:250,000-scale paper map, the minimum distance that can be represented is about 125 m. On a 1:1,000,000-scale paper map, the resolution is 500 m.

For example, in an assessment of Nevada's mineral resources (Cox et al., 1996), the tract permissive for pluton-related deposits was defined as an area extending 10 km outward from the outcrop of a pluton or, in the case where the pluton had a geophysical expression, from the inferred subsurface boundary of the pluton based on its geophysical expression. It also includes plutons whose presence was inferred from geophysics or from the occurrence of skarn mineralization. Some pluton-related deposit types such as skarns are known to occur less than about 2 km from the pluton contact; however, the authors could not be portray this more appropriate boundary at the published 1:1,000,000 scale of the map because they would appear as narrow slivers scattered throughout the map.

Cover

Years ago when the authors were doing research for their Ph.D. degrees, they independently discovered that the most useful geologic predictor of metal production was the percentage of cover, such as alluvium. The more cover, the lower historic metal production. The reason for the relationship has more to do with difficulty of exploring under cover than it does the existence of undiscovered metal-bearing deposits under cover. Undiscovered mineral deposits commonly occur beneath some sort of cover (e.g., alluvium). Therefore, to be meaningful, an assessment must be able to predict what might be present in covered areas. A geologic map is the primary local source of information for delineating domains and identifying which domains are permissive for different deposit types. This presents a challenge to assessors because, for most areas, geologic maps of the rocks beneath cover are not available.

Regional or metallogenic settings of mineral deposits can provide guides on general locations of deposits under cover if the geology can be projected under cover (figure 7.4). Fortunately, some scientists have experience in making geologic maps under cover. The Australians have had to deal with the problem of cover for some years (Wilford, 2000). As might be guessed, however, the methods require detailed geophysics, which, because of its expense, can be applied only in relatively small areas. One area of active research that should prove useful in locating possible places of mineralization under cover is structural geology. Many faults can be identified under cover with geophysics—some kinds of faults are related to certain types of mineralization such as porphyry copper deposits (Berger, Drew, and Singer, 1999).

About 50 percent of the surface is covered by apparently barren rocks and sediments in many large regions, such as Australia, the United States, and parts of Europe. Because the majority of mineral deposits exposed at the surface are believed to have already been found, a prime concern of assessments in such cases is the nature of and depth to possible mineralized systems under this cover. Resource assessments of areas with resources under cover must rely on extrapolation from surrounding areas, new geologic maps of the rocks under cover, or analogy with other well-explored areas that can be considered training tracts.

Cover has a profound effect on methods and procedures of conducting assessments. Known deposits and occurrences are present on the margins of covered tracts in fortunate cases. Geology under cover is seldom known. Geophysical methods typically have responses that are variably attenuated depending upon thickness of cover—geochemical responses are even more attenuated. Many powerful assessment methods that have been developed in recent years have been based on relationships of geochemical and geophysical variables to deposits learned from exposed deposits. In order to use these methods in covered regions, the relationships need to be reexamined based on conditions under cover—not an easy task because of sparse information on deposits under cover.

Geographic Information System Delineation

With the power of modern computers and digital data sets, there is considerable interest in using GIS technology to delineate permissive tracts or to identify possible targets. Where permissive rocks are exposed, this technology allows rapid and precise delineation. The use of GIS techniques has been recommended as reproducible, objective, accurate and more rapid than expert approaches (Raines and Mihalasky, 2002). Where permissive rocks are covered by younger geologic units such as sediments or volcanic ash or by tectonically emplaced cover, these claims are questionable.

Typically in a GIS-based approach, delineations under cover are dealt with by using a buffer around the exposed permissive rock (Ludington et al., 2008). The buffer is commonly described as the optimum prediction distance where the distance has the maximum association (called maximum contrast) of known mineral deposits with the layers being considered in the training data (Carranza and Hale, 2002; Raines and Mihalasky, 2002). When these contrasts have the strongest confidence levels, it is argued that an optimal distance for prediction of the occurrence of mineral deposits has been determined. The problem is that the number of discovered mineral deposits diminishes as one moves away from exposed bedrock because of either the absence of a permissive geologic setting for the deposit type or, more commonly, the lack of sufficient and effective exploration as one moves away from exposed bedrock to covered bedrock in most places. Thus, the number of mineral deposits that have been discovered under cover is clearly a biased sample of the number of existing deposits under cover in a permissive geologic setting, and "optimum prediction distance" is a misnomer.

Unless the geologically permissive rocks are specifically mapped using interpolated geology and geophysics, the basis for the delineation under the cover violates the principals of delineating permissive tracts. For example, if a buffer zone around the exposed permissive rocks were used to delineate the permissive tract under cover in figure 7.4, the extent of covered permissive rock would have been incorrectly estimated, resulting in poor estimation of the number of deposits under cover, and it also would have completely missed the part of New South Wales that is permissive for covered porphyry copper deposits. Thus, decision-makers would be misinformed about both locations and amounts of mineral resources, leading to misallocation of human and financial resources. The advantage of the GIS techniques being more rapid than expert approaches to delineation under cover is of questionable value if the results are biased and inconsistent with the other parts of the assessment. Using GIS buffers to delineate permissive rocks under cover results in tracts that are inconsistent with true permissive rock boundaries and are inconsistent with other parts of three-part assessments, such as matching the geologic setting of the deposits type and estimating the number of deposit using deposit-density models (Singer, 2008). The quickest way off of a mountain might be jumping off a cliff, but that is probably not the best way to get down.

Subdivision of Tracts

If an area being assessed has different scale geologic maps or different quality maps, it may be desirable to delineate separate tracts to represent different qualities or amounts of information. Similarly, if a tract has parts where there is reason to believe that deposits are more or less likely to exist than in other parts of the tract, the tract should be divided into subtracts. Ideally, the probability of the existence of a deposit would be the same everywhere within a tract, and a permissive tract should be subdivided to try to accomplish this equal probability state. Because probabilities of deposit occurrence are not commonly estimated within tracts, a more practical guide would be that tracts should be divided whenever the expected number of deposits or the level of uncertainty varies within a tract. Having the same probability of deposit occurrence with a tract makes it easier to statistically divide or recombine tracts and associated estimated number of deposits, if that ever became necessary.

Concluding Remarks

Delineating tracts that are permissive for undiscovered deposits relies on linking of geologic settings of deposit types as identified in models to geologic environments as interpreted from maps. Deposit models play the central role in identifying relevant information and in integrating the various kinds of information to delineate permissive tracts. Map scales used in preparing mineral deposit models and in delineation of permissive tracts therefore are central to properly prepared assessments. When assessing undiscovered mineral deposits by type, the base map selected is typically affected by the availability of only a small number of geologic map scales of the area, publication scale of the assessment, and time and space limits on the assessors. A frequent effect of these limits is that the map scale selected is not ideal, because the delineated permissive tract containing the geologic units that could host the deposits may also include unreported units that could not contain the deposits. The delineated tract may also contain unreported geologic units that cover the geologic units of interest or unreported parts that are too deeply buried to be permissive. In these situations, areas of delineated tracts are larger than necessary due to inflation by unaccounted for nonpermissive areas, or by covered areas that are poorly explored (Singer and Menzie, 2008).

Because of reliance of these assessments on the link between deposit types and geology, assessing mineral resources under cover has a great deal of uncertainty in delineated boundaries and in associated estimates of number of deposits. For this reason, considerable effort is warranted in mapping the geology under cover. Better quality and detail in maps also can significantly reduce the risk of exploration failure as discussed in chapter 10.

8

Estimating the Number of Undiscovered Deposits

Perspective

The third part of three-part assessments is the estimate of some fixed but unknown number of deposits of each type that exist in the delineated tracts. Until the area being considered is thoroughly and extensively drilled, this fixed number of undiscovered deposits, which could be any number including 0, will not be known with certainty. This number of deposits has meaning only in terms of a grade-and-tonnage model. If this requirement did not exist, any wisp of minerals could be considered worthy of estimation, and even in small regions, we would need to estimate millions of "deposits." For example, it is not difficult to imagine tens of thousands of fist-sized skarn copper "deposits" in parts of western United States—even in this example, we have used "deposit" size to provide important information. In another example, Wilson et al. (1996) estimated five or more epithermal gold vein deposits at the 90 percent level but provided no grade-and-tonnage model, so these estimated deposits could be any size. To provide critical information to decision-makers, the grade-and-tonnage model is key, and the estimated number of deposits that might exist must be from the grade-and-tonnage frequency distributions.

In three-part assessments, the parts and estimates are internally consistent in that delineated tracts are consistent with descriptive models, grade-and-tonnage models are consistent with descriptive models and with known deposits in the area, and estimates of number of deposits are consistent with grade-and-tonnage models. Considerable care must be exercised in quantitative resource assessments to prevent the introduction of biased estimates of undiscovered resources. Biases can be introduced into these estimates either by a flawed grade-and-tonnage model or by the lack of consistency of the grade-and-tonnage model with the number-of-deposit estimates. For this reason, consistency of estimates of number of deposits with the grade-and-tonnage models is the most important guideline. Issues about consistency of mineral deposit models are discussed in chapters 3 through 6. Grade-and-tonnage models (chapter 6), which are the first part of three-part assessments, are of particular concern. In this chapter, the focus is on making unbiased estimates of the number of undiscovered deposits.

Estimates of number of deposits explicitly represent the probability (or degree of belief) that some fixed but unknown number of undiscovered deposits exist in the delineated tracts. As such, these estimates reflect both the uncertainty of what may exist and a measure of the favorability of the existence of the deposit type. Uncertainty is shown by the spread of the number-of-deposit estimates (quantiles) associated with the 90 percent quantile to the 10 or 1 percent quantile (figure 8.1)—a large difference in the numbers suggests great uncertainty. Favorability can be represented by

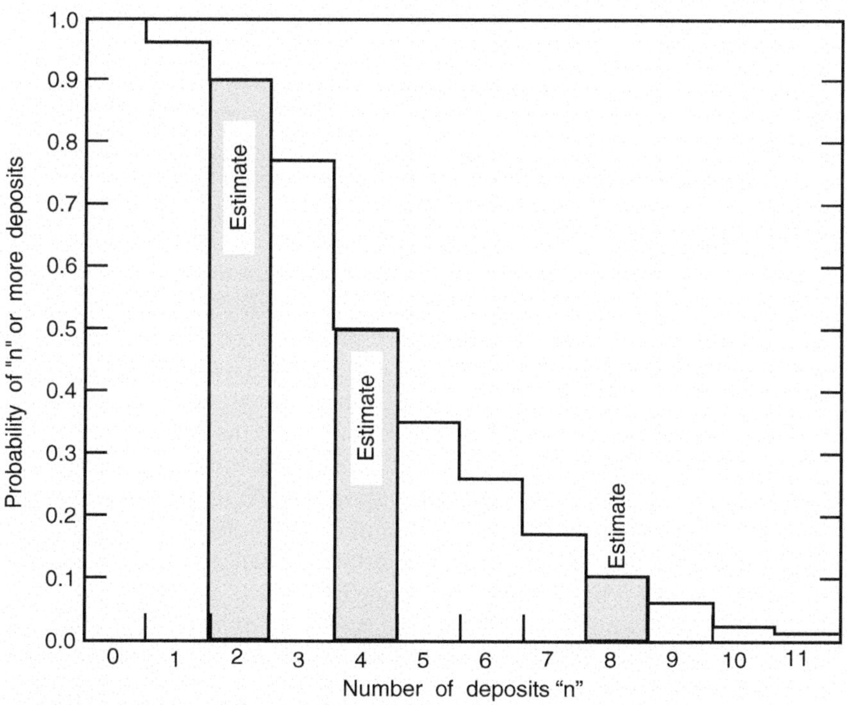

Figure 8.1 Three estimates of "n" or more deposits (shaded) along with interpolated values.

the estimated number of deposits associated with a given probability level or by the expected (mean) number of deposits.

Why Estimate Quantiles?

In the first modern quantitative resource assessment, Allais (1957) used a Poisson distribution to model the occurrence of mineral deposits—justification for its use was an unpublished study of the number of mineral deposits exceeding a certain value in eighty-nine French administrative divisions. The Poisson distribution has several advantages over some alternative discrete distributions, such as requiring estimation of only one parameter and simple calculation of the probability of individual numbers of deposits (figures 8.2, 8.3). A Poisson distribution is also easy to work with mathematically. Among the properties of the Poisson distribution are that the mean is equal to the variance, and each sample outcome is independent from sample to sample. Thus, if a mineral deposit exists in a cell, the probability that the adjacent cell contains a deposit is not affected. A number of studies of deposit density suggest that this independence assumption may not be appropriate—that is, deposits seem to be clustered in space (see summary in Harris, 1984). Negative binomial distributions seem to fit the observed distributions of deposits more commonly than do Poisson distributions. Negative binomial distributions require estimation of two parameters, have variances that are greater than their means, and have the property that knowing a cell contains a deposit would change the estimate that the adjacent cell contains a deposit.

When assessors are not constrained by a specific probability distribution, the estimates do not always fit distributions that assume a constant probability such as the Poisson distribution. For example, about half of the estimates of numbers of copper-bearing deposit types that were made in an Alaskan assessment (Eberlein and Menzie, 1978; Grybeck and DeYoung, 1978; MacKevett, Singer, and Holloway, 1978) were found to be consistent with Poisson distribution, but about half were not and required a negative binominal distribution (Harris, 1984). Griffiths and Ondrick (1970) demonstrated with simulation that even where a Poisson distribution fits the original frequency of deposits, introducing cover over part of the area induces clustering of deposits, rejection of the Poisson distribution, and fit of the observed distribution of deposits to the negative binomial distribution. A histogram of a negative binominal distribution looks different than one from a Poisson distribution in that it has a higher frequency of deposits on the high number of deposits (right-hand) side. That is, the right side of a negative binomial distribution has more probability than does a Poisson distribution (heavier tails). Estimation of numbers of deposits should allow for this possibility.

Number-of-deposit estimates can be made using in any of several forms. One form is to elicit the entire frequency distribution that describes the

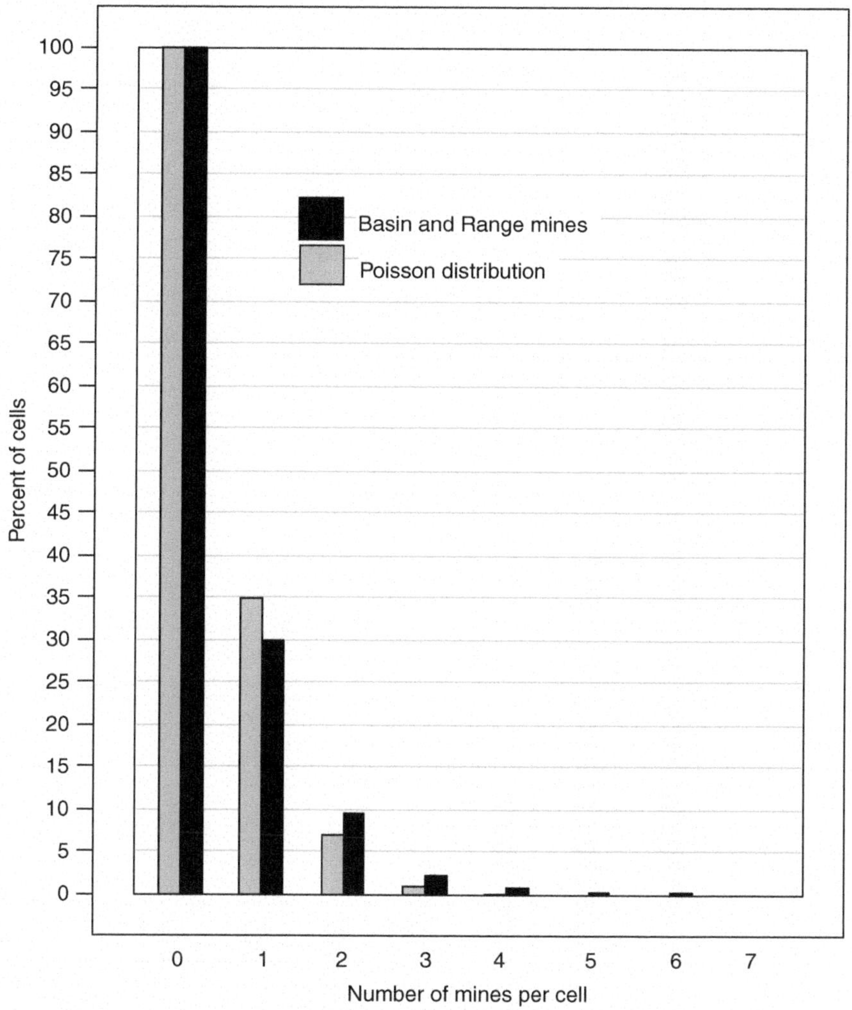

Figure 8.2 Distribution of number of mines in cells in the Basin and Range mines compared to the number in a Poisson distribution with the same mean (154 mines, 357 cells). After Slichter (1960).

probability of the tract containing a given number of deposits. This method can be used if only a few deposits are expected to occur in the tract but is cumbersome if the estimated number of deposits is large or has a high uncertainty. A second form is to use a triangular distribution to estimate the entire probability distribution by eliciting estimates of the lowest, highest, and most common number of deposits. If the distribution is nearly symmetrical, a triangular distribution can give a satisfactory representation of the distribution of undiscovered deposits. However, the triangular distribution will

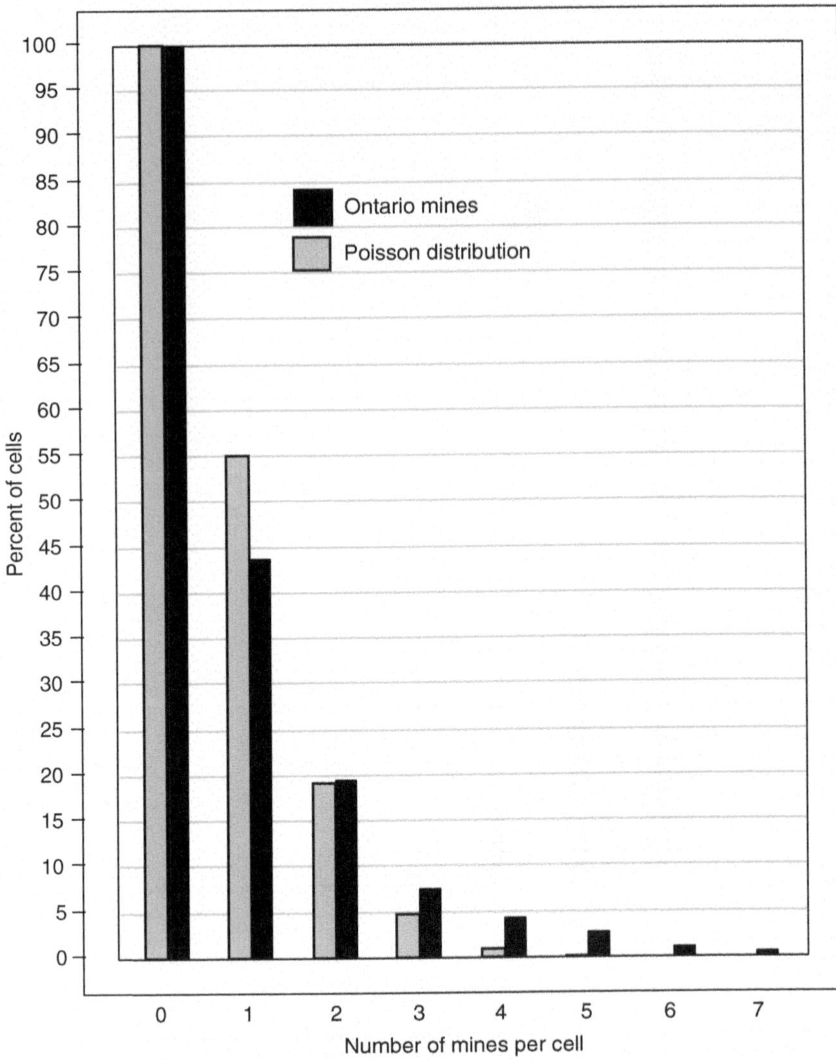

Figure 8.3 Distribution of number of mines in cells in Ontario compared to the number in a Poisson distribution with the same mean (147 mines, 185 cells). After Slichter (1960).

lead to biased estimates if the real probability distribution is highly skewed. The most commonly used method of estimating undiscovered deposits is to elicit an estimate of a number of deposits that the tract will equal or exceed at given probabilities. Commonly, estimates are made in quantile form. That is, numbers of deposits are elicited that estimate the number of deposits or more that may be present in the tract at a 90 percent level, a 50 percent level, and a 10 percent level. If necessary, estimates can also be made at a

5 or 1 percent level. Interpretation of these quantile estimates requires care. Because of their format where the number of deposits is represented by at least so many deposits (Root, Menzie, and Scott, 1992), a unique underlying probability distribution cannot be specified on the basis of the elicited distribution alone. Root, Menzie, and Scott (1992) provide a methodology for obtaining a probability distribution that is consistent with these quantile estimates. The probability distribution is required if one wants to use Monte Carlo simulation to properly combine estimated numbers of deposits with grade-and-tonnage models to estimate total undiscovered resources and their associated uncertainty in the tract.

Whether commonly observed deposit clustering is caused by natural clustering of deposits or is a consequence of the interaction of how we sample the distribution of deposits, a frequency distribution with heavier tails than a Poisson must be expected and allowed in a quantitative assessment. All evidence available in hundreds of estimates of number of deposits in tens of three-part assessments suggests that a significant proportion of estimates of number of deposits should have some skewness toward the right and that negative skewness of estimates is quite unlikely (see, e.g., figure 4.2). For all of these reasons, in three-part assessments number-of-deposit estimates are made for the number of deposits associated with the 90, 50, 10, 5, or 1 percent quantile without the requirement that the estimates fit any specific frequency distribution.

Effects of Subdividing Tracts

We recommend in chapter 7 on delineation that during an assessment, a tract should be subdivided whenever different degrees of uncertainty or different likelihoods of containing deposits can be identified within the tract. For each of the new tracts, separate estimates of the number of undiscovered deposits should be made. After the assessment is complete, it is not uncommon that there are requests to subdivide tracts and associated estimates of number of deposits. In situations where a Poisson distribution fits the estimated number of deposits for a tract, such as in the example in figure 8.4, the constant probability and the mathematical properties of the distribution allow relatively easy subdivision of the estimates into new estimates that have means proportional to the areas of the new subtracts, as shown in figure 8.5. This is accomplished by first estimating the expected number of deposits using the following equation (Singer and Menzie, 2005):

$$E(n) = 0.233\,(N_{90}) + 0.4\,(N_{50}) + 0.225\,(N_{10}) + 0.045\,(N_{05}) + 0.03\,(N_{01}), \quad (8.1)$$

where N_x = the estimated number of deposits at the xth percentile. If only the 90th, 50th, and 10th percentiles have been estimated, use the 10th estimate for the 5th and 1st also. This equation closely approximates the expected

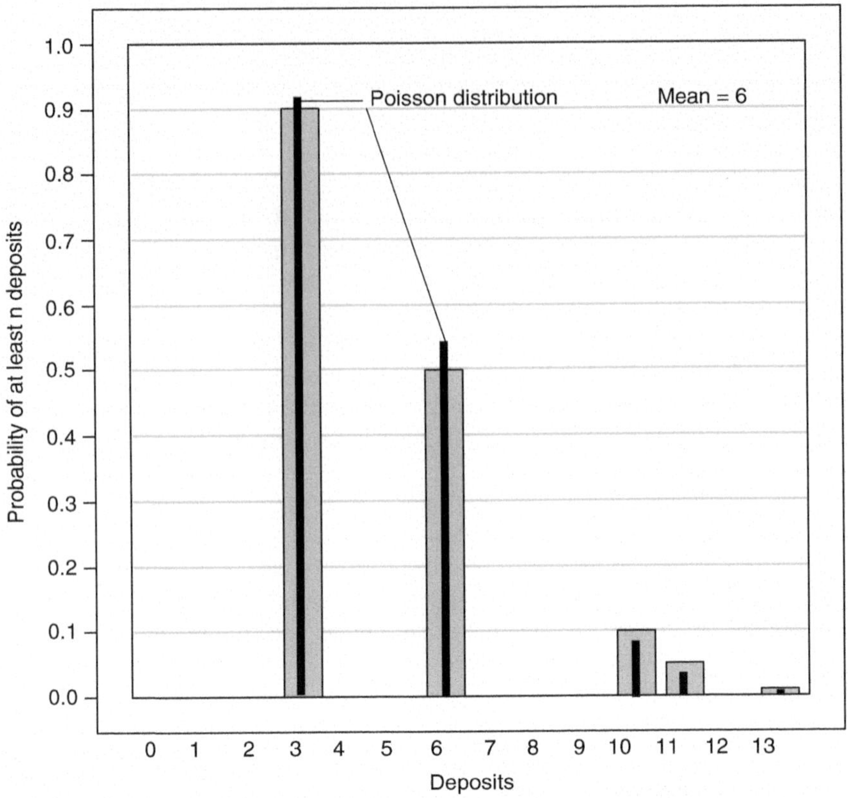

Figure 8.4 Estimated number of deposits or more that is consistent with a Poisson distribution that has a mean of six deposits.

number of deposits estimated in the MARK3 simulation program (Root, Menzie, and Scott, 1992). The expected number of deposits is then used as the mean (λ) in a Poisson distribution to generate cumulative probabilities to compare with the estimates in the original tract. If the number of deposits associated with the 90th, 50th, 10th, 5th, and 1st percentiles of the Poisson distribution closely match those of the tract estimates, the tract can be subdivided with each new mean (λ) of a subtract proportional to its relative area; that is, the expected number (E(n)) from equation 8.1 is divided by the proportional area of each subtract, and the new values are estimates of the tract's mean, which becomes the mean (λ) in a Poisson distribution. This new mean can then be used with the Poisson distribution to generate the estimates at the appropriate percentiles. However, when the negative binomial distribution or some other distribution that has heavier tails than the Poisson distribution fits the estimated number of deposits, there is no tractable way to subdivide the tract and have the estimated number of deposits reflect both the expected number of deposits and the uncertainty of the estimate.

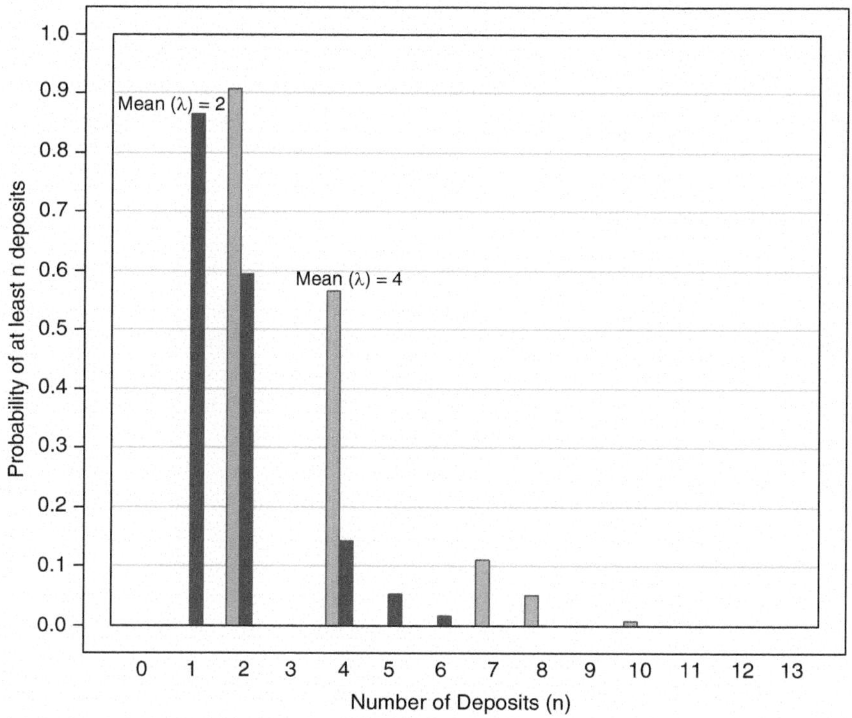

Figure 8.5 Estimated number of deposits or more for two tracts representing one-third and two-thirds of the area of a tract where number of deposits is distributed as a Poisson distribution with mean (λ) = 6.

Estimating Number of Deposits

This section briefly examines (1) factors that influence estimation of undiscovered deposits, (2) development of methods to estimates numbers of undiscovered deposits, and (3) recent U.S. Geological Survey (USGS) practice in estimating numbers of undiscovered deposits.

The first factor that influences the estimation of undiscovered deposits is the intended use of the estimate. Because the three-part method uses grade-and-tonnage models to estimate the mineral endowment of areas, it is important that estimated undiscovered deposits be compatible with the grade-and-tonnage models. That is, one should expect 80 percent of the estimated deposits to lie between the 90 and 10 percentile estimates of the tonnages (or grades) depicted in the appropriate model and half of the estimated deposits to exceed the median of the distribution.

A second factor that influences the estimation of undiscovered deposits is the available geologic data: its amount, type, and nature. For example, different types of geoscience data vary in the effectiveness with which they reflect different types of deposits. In addition, the density and spatial distribution of sampling associated with the survey affect the estimation of undiscovered deposits. Further, one must consider the nature of the responses themselves. Are a number of anomalous samples present, or are values within background levels? Finally, one must combine different types of geoscience data to develop estimates. What constitutes an appropriate combination of data varies with deposit type, and the logic for combining different data may also vary with deposit type.

A third factor that influences estimation of undiscovered deposits is the mineral exploration that has taken place in the area: its type, amount, spatial distribution, and effectiveness. In extreme cases where exploration has been thorough, estimates of undiscovered deposits may be reduced to zero. More commonly, exploration is only partial; in such cases, the effects on undiscovered resource estimates may be difficult to interpret. One effect of exploration that has been suggested by many authors, including Singer and Mosier (1981b), is that the previous exploration may have discovered the larger deposits, and remaining, undiscovered deposits may be, on average, smaller than those already found. In such cases, undiscovered deposits would not be comparable to those in grade-and-tonnage models.

In the assessment, an important factor affecting estimates of number of deposits is the proper distinction between known deposits and prospects. Deposits referred to as "discovered" or "known" are reported in the published literature to be well explored in three dimensions and not open in any part and to have published tonnages and grades. Explored metal occurrences not meeting these criteria are classified as prospects even if they are being mined if there is an indication that more resources are expected. Such a definition is necessary to avoid either double counting or missing some resources. Some of these prospects typically represent some of the undiscovered deposits estimated.

Historically, two end-member approaches to estimate numbers of undiscovered deposits have been used or recommended. The first approach uses traditional statistical methods, including discrete frequency distributions, and multivariate statistical methods. The second approach uses expert judgment to estimate the numbers of undiscovered deposits that may occur in delineated tracts. While application of traditional statistical method is appealing because it seems to offer a rigorous, objective way to estimate undiscovered resources, in practice applications of traditional statistical methods have not been very satisfying except for the deposit density models. Most models are not very flexible with regard to the amounts and kinds of information they use. Commonly, mineral assessments are needed for regions with highly variable geoscience information, which makes the traditional statistical methods less useful.

What Is Being Estimated?

In order to capture all resources from known deposit types, it is necessary to have some rules about how different categories of resources are counted.

The concern is that everything is counted (or estimated) once, but only once. In general, we simply need to estimate the number of undiscovered deposits because the discovered deposits are already counted. In practice there are some gray areas that need to be resolved. Where a deposit is thoroughly drilled but the information has not been made public, the deposit should be counted as undiscovered (with high certainty). Mineral deposits that are only partially explored are counted as undiscovered if assessors believe that the deposits would, upon further drilling, be from the same distribution of sizes and grades as represented in the grade-and-tonnage model—this may seem strange if the partial resource estimates are well established. The problem is that if deposits in this situation were counted as known deposits, the grade-and-tonnage model would not be appropriate, and any resources added through time to these deposits would not accounted for in any category. These kinds of situations help demonstrate the value of using care in performing the assessment.

Can Experts Estimate Number of Deposits: An Experiment

When geologists place on a map an inferred contact between two rock units that has not been seen, subjective judgment is used. The same is true when tectonic boundaries are placed on maps in large regions. Every day each of us makes subjective judgments, and there is strong evidence that a Bayesian probability process is fundamental in learning (Kording and Wolpert, 2004).

Subjective probabilities such as used here variously have been called degrees of belief or propositional probabilities. The oldest and probably most commonly practiced form of subjective estimation is gambling. For example, Stern (1991) showed that the distribution of actual margins of victory versus predicted point spreads in National Football League games has a mean of zero, indicating that nonscientists can make unbiased subjective estimates. Geologists commonly make similar estimates that, although not explicitly quantitative, are subjective and have uncertainty, such as making geologic cross sections. Examples from different fields of study (Murphy and Winkler, 1984; Stern, 1991) demonstrate that under some conditions subjective estimates can be unbiased and reliable. For example, figure 8.6 plots more than 150,000 subjective estimates made by meteorologists of the probabilities of precipitation against the observed relative frequency of precipitation. For every forecast probability except the 100 percent estimates, the estimates fall on the expected 1:1 slope, demonstrating unbiased estimates.

Use of expert judgment to estimate the number of undiscovered deposits within tracts has caused concern by some geologists about the "accuracy" of the assessments (Harris and Rieber, 1993). Usually critics call for a demonstration of a place where an assessment has predicted the occurrence of a mineral deposit in advance of exploration. While such examples

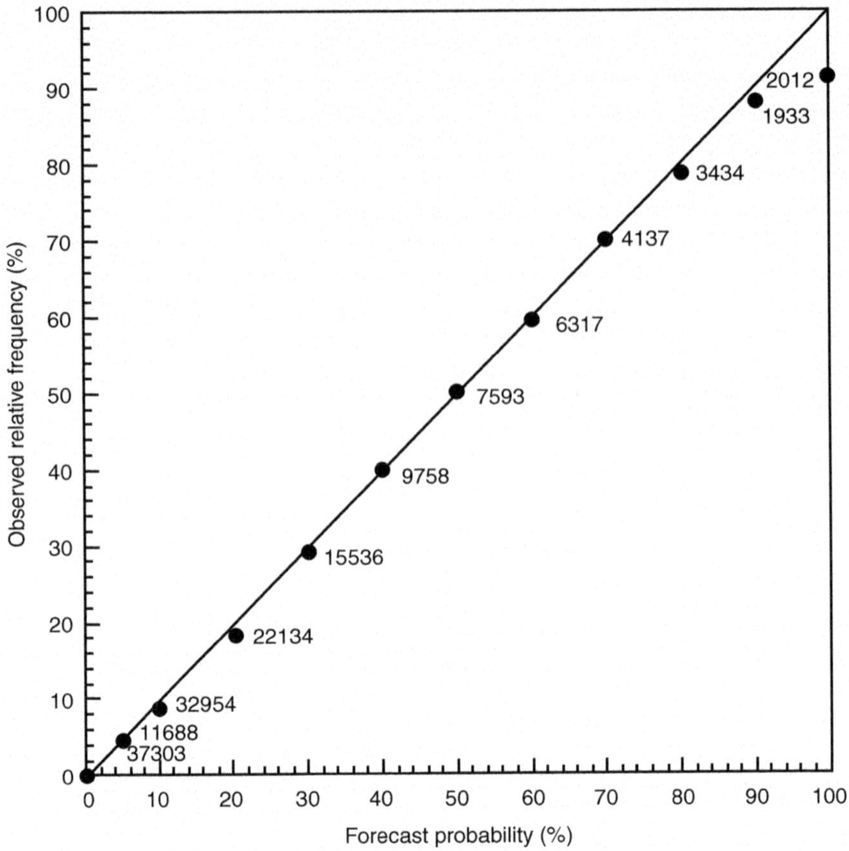

Figure 8.6 Comparison of subjectively estimated chance of precipitation and observed frequency. Numbers near symbols represent number of estimates at that forecast probability. The straight line represents no bias. After Murphy and Winkler (1984).

impressively document assessment successes such as the discovery of Pebble Copper in Alaska after a published estimate (MacKevett, Singer, and Holloway, 1978), they may not provide the best basis for evaluating a mineral assessment.

To test whether geologists could make unbiased estimates of the number of deposits, an experiment was conducted with well-explored areas to compare geologists' estimates with the number of known deposits. A pool of thirteen areas was developed to mitigate possible recognition of individual areas by the geologists. In the experiment, ten areas and their data sets were randomly assigned to each of twenty-four geologists. This resulted in a total of 239 possible experiment sets (one geologist only had nine areas). In general, recognition of areas was not a problem because the data sets had all geographic names and coordinates removed and were prepared in odd shapes and sizes that tend to reduce recognition. In cases where geologists thought they recognized areas, they were asked to record their guess and proceed with the experiment as though they did not "know" if they were correct. In most cases, the geologist's guesses proved erroneous. In a few cases, geologists recognized one of the areas and another area was substituted for the compromised area. In one case a geologist recognized four areas, and thus only nine measurements were available for that geologist.

The areas were from throughout the North American Cordillera, all from geologic settings that, considered broadly, contain one or more porphyry copper deposits, which was the deposit type estimated. Some of the specific areas are believed to be barren of porphyry copper deposits; others contain one or more deposit. Areas barren of porphyry copper deposits may host other types of mineral deposits. The areas ranged in size from 57 to 1,210 km^2 (22–468 square miles); the median size of the thirteen areas is 342 km^2 (132 square miles).

Data available for the areas include geologic maps, stream sediment geochemical samples, aeromagnetic data, and mineral occurrence data. All data sets contain geologic maps that vary in scale from 1:48,000 to 1:250,000. Two maps are 1:48,000, seven are 1:62,500, one is 1:63,360, and three are 1:250,000. All maps show important lithologic and structural units. Smaller scale maps tend to show only the boundaries and lithologies of igneous units and larger structural features; larger scale maps show phases and compositions of individual intrusives and show more detailed structural features. Only one map shows hydrothermally altered rocks, but two others show locations where sulfide minerals were encountered during field mapping. All data sets also contain stream sediment geochemistry data. Density of sampling varies from 0.031 samples to 0.48 samples/km^2. Geochemical data are 33-element semiquantitative spectrographic determinations. The sample media were generally the 80-mesh fraction of stream sediments. Nine of the thirteen data sets contain aeromagnetic data; flight line spacing varied from 0.8 to 8 km (0.5–5 miles) spacing. Eleven of the thirteen areas have mineral occurrence maps and short descriptions of the occurrences. Locations and descriptions of porphyry copper deposits were removed from the data set, but associated deposit types such as vein and skarn deposits were left in the sets.

Because the size of the thirteen areas was limited, the number of porphyry copper deposits present ranged from 0 to 2 deposits. To determine how well each geologist estimated the number of deposits, the expected (mean) number estimated by that geologist for each area was plotted against the observed number of deposits. Thus, if a geologist perfectly estimated the number of deposits in each of ten areas, the plotted points would fall on x,y-coordinates of 0,0, 1,1, and 2,2, forming a straight line with a slope of 1.0, and the correlation coefficient would be 1.0. The more the slope and the correlation coefficient deviated from 1.0, the worse the geologist was at estimating the number of porphyry copper deposits. A reasonable test is if the correlation coefficient is significantly greater than zero, then the geologist can be considered to have demonstrated an ability to estimate the number of deposits. In table 8.1, we can see that some geologists were excellent at estimating number of porphyry copper deposits, and others were not as good. This experiment demonstrates that not all geologists should be considered experts at estimating number of mineral deposits. It also shows that some geologists are capable of making unbiased estimates of the number of undiscovered mineral deposits. It is also possible that, with some training and calibration, those geologists who did not do as well at estimating porphyry copper deposits might become trusted experts.

Can these individual estimates be improved upon? In many resources assessments, a group of experts make independent estimates and then make consensus estimates, as discussed below. In the experiment testing

Table 8.1. Results of twenty-four geologists' estimates of number of porphyry copper deposits.

Geologist	r	df	b_2
1	0.785**	8	1.54
2	0.771**	8	1.04
3	0.887**	8	1.71
4	0.686*	8	0.986
5	0.382	8	0.544
6	0.851**	8	4.79
7	0.685*	8	0.772
8	0.338	8	0.458
9	0.550	8	0.903
10	0.044	8	0.058
11	0.741*	8	2.11
12	0.006	8	0.063
13	0.627	8	0.354
14	0.082	8	0.154
15	0.611	8	0.739
16	0.795**	8	0.718
17	0.819**	8	1.27
18	−0.138	8	−0.059
19	0.744**	8	0.948
20	0.639	7	1.55
21	0.336	8	0.514
22	0.398	8	1.21
23	0.723*	8	1.95
24	0.569	8	0.807

r = correlation coefficient, df = degrees of freedom, b_2 = slope of regression line.
* = significant at 5% level, ** = significant at 1% level.

geologists for their ability to make estimates, the geologists were never brought together because the experiment was carried out over several years and different parts of the country. However, we can test to see if the average of the group improves the estimates. In figure 8.7 the expected value (average) of the estimates of geologists is plotted against the observed number of deposits for the thirteen tracts. Only the eleven geologists who demonstrated an ability to estimate porphyry copper deposits by having estimates significant at the 5 or 1 percent level were used in this plot. The correlation coefficient is 0.85, which is highly significant, and the slope of the regression line is 1.58, which is not significantly different than the ideal value of 1.0. From this we conclude that groups bring some stability to the estimation process; that is, they can reduce the variability of the estimates. Interestingly, when all twenty-four of the geologist were included is this kind of plot, the correlation only decreased to 0.74, which is significantly different than zero, and the slope changed to 1.7 which is also not significantly different than 1.0. Thus in this experiment with twenty-four estimators, the power of the group made up for the less able estimators. In actual assessments with smaller number of estimators, it is not clear that one should rely on a few able estimators to cover for estimators who are not able.

The decades of experience of subjective and objective forecasting in meteorology provide insight into how the process of making subjective assessments in mineral resources might be improved. Murphy and Winkler (1984) found that consensus schemes performed better than almost all individual forecasters and that the best forecasts were made when objective forecasts were part of the information supplied to subjective forecasters. Among their recommendations were more effective use of many information sources,

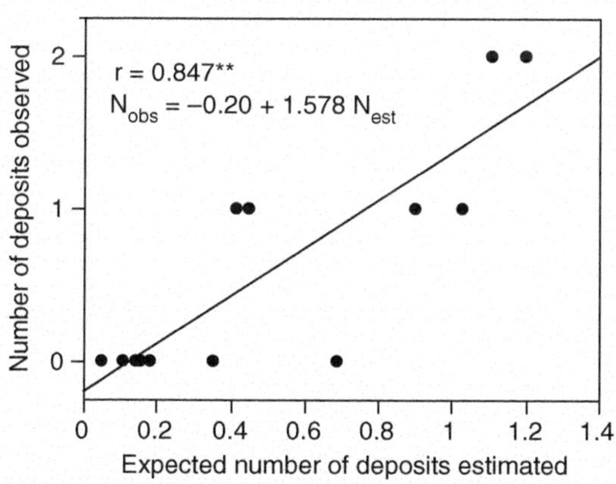

Figure 8.7 Comparison of estimated number (N_{est}) of porphyry copper deposits to number of deposits known (N_{obs}) in blind test of thirteen tracts by eleven geologists designated as experts. r = correlation coefficient. **Correlation significantly different than 0.0 at the 1 percent level.

motivation to encourage forecasters to improve their performance, provision of formal procedures to assist forecasters in quantifying their uncertainty in terms of probability, and quick and extensive feedback concerning performance. Quick and extensive feedback might be difficult to apply in mineral resource assessments, except possibly through training exercises such as the one discussed above. The emphasis here on subjective estimation and the use of objective guidelines stems from our belief that few objective quantitative methods have yet to be shown to be consistently effective and unbiased in estimating the number of undiscovered deposits. Even the powerful mineral deposit-density models should not be relied upon in all cases (see discussion in chapter 4). It is important to note that three-part assessments are a form of product, not a method, and therefore do not preclude the use of any method that is consistent with the other parts of the assessment. We should always use the best possible methods of making quantitative assessments.

Estimation Process

The goal in making number of undiscovered deposit estimates is to make unbiased estimates of the number with a method(s) that produces the minimum variance. Commonly, expert judgment is used because of the high uncertainty of such estimates and, more important, the data upon which such estimates must be made typically represent a mixture uneven in coverage, types, and quality. A formal elicitation process is used in that particular criteria are used, experts are selected, the method is designed, and the response mode is specified (Meyer and Booker, 2001). The general approach used by the USGS in estimating numbers of undiscovered deposits in three-part assessments has evolved on the basis of experience in earlier assessments and the results of experiments to test the ability of experts to estimate undiscovered deposits. Some of the changes the USGS has adopted in using experts to estimate undiscovered deposits have included use of digital information in mineral deposit models, use of teams of experts to estimate undiscovered deposits, calibration of estimators through the use of formal training sets or through apprenticeship, formalization of the process by which estimates are elicited, introduction of statistical guides such as deposit densities to assist experts, and development of mechanisms that feedback implications of their estimates to the estimators.

Commonly, the USGS has used teams of estimators that have included experts in the geology of the region being assessed, economic geologists who are experts on the deposit type(s) being assessed, and resource analysts who are experienced with the resource assessment process. Only team members who are comfortable with making estimates are asked to do so. Team members who are not comfortable with the estimation process act as consultants in their area of expertise. Individual team members study the available geoscience data for the region to be assessed before attending the assessment session. The resource analyst knowledgeable about statistics will test the area's known deposits to determine if the global grade-and-tonnage model(s) are significantly different than the local deposits that have been drilled in three dimensions (see chapter 6). It is not enough to note that the known deposits fall on the published grade-and-tonnage model because all of the known deposits might have tonnages that are on the low tonnage part of the model. The purpose of grade-and-tonnage models is to provide unbiased representations of the grades and tonnages of undiscovered mineral deposits in a tract. Tonnages and grades of the known deposits in an area being assessed should always be tested against the model before the assessment. If there are significant differences, new grade-and-tonnage models might need to be developed.

The assessment session begins with an overview by the leader of the purpose, goals, and plans for the meeting. A short overview on the properties of statistics of quantiles is useful at this point also. This is followed by presentation of the available geoscience data for the region by the regional experts. Once the types of deposits that might occur in the region have been identified and tracts have been delineated (see chapter 7) that might contain undiscovered deposits of these types, then assessment proceeds to estimation of the number of undiscovered deposits. Before the estimation, any appropriate guides such as deposit densities should be presented to the estimators. In addition, the estimators would be reminded that half of each number-of-deposit estimate should exceed the deposit type's median tonnage—this is an excellent way to identify possible biased estimates or misunderstandings about what is being estimated.

Two general strategies tend to be used for estimation. The first relies on a comparison of regions geologically similar to the one being evaluated. In this approach, the geologist compares relevant features of the region with those of other well-explored regions, and after evaluation of these features and the sizes of the regions, the expert makes subjective estimates of the number of deposits in the domain at fixed probability levels. This approach, which relies on the expert's mental frequency distribution of deposits in regions with similar characteristics and areas, may be used with both regional-scale and more detailed data. Mineral deposit densities, which were discussed in chapter 4, are useful decision aids that can help in using this method of estimation. However, when mineral deposit densities are used to estimate numbers of undiscovered deposits, the estimates must be adjusted to reflect any risk that the tract being evaluated is barren. This is done by multiplying each estimate by one minus the probability that the region is barren, that is, the probability that the tract has zero deposits of the type being estimated.

The second approach relies on identifying potential exploration targets for the type of deposit being estimated. This requires detailed information. In this case, individual targets can be compared to descriptive deposit models to evaluate the likelihood that they are deposits of the type being estimated. If the method of target identification is used as the geologic basis for estimation, the frequency distribution must be formed from probabilities assigned to individual targets. For example, a region may contain three targets. One may exhibit many features associated with a particular deposit type but may not have been sufficiently explored to establish if it is, in fact, a deposit. The expert may believe that the probability of this target being a deposit is very high and assign it a value of 0.9. A second target may have some features of known deposits but may differ in some important aspect; the expert might assign a probability of 0.3 to this target. The third target may only be inferred on the basis of geophysical properties of the deposit or associated rocks. In this case, the expert would usually assign the target a low probability (0.05) of being a deposit. Cox (1993) provides one example of this method. If the probabilities that individual targets are deposits are mutually independent, the probability distribution of deposits in the domain can be calculated explicitly. The probability that the domain contains 0, 1, 2, and 3 deposits is 0.0665, 0.6305, 0.2895, and 0.0135, respectively. This frequency distribution can be combined to give the probability that the domain contains at least 0, 1, 2, and 3 deposits; these probabilities are 1.0, 0.9335, 0.303, and 0.0135, respectively.

After each geologist has made their initial estimates of the numbers of undiscovered deposits present in the tract, the results are recorded and examined. The results are analyzed for the possibility of bias. The analysis focuses on estimates that are higher than the highest deposit-density estimate for the type, estimates that imply more metal in the tract than the amount know in the world, and estimates that either underestimate uncertainty or are skewed toward the right. If there is a large difference between the estimates by different scientists, a discussion takes place focusing on what evidence led the estimators to their estimates. Attention is paid to the high and low estimates. The purpose of the discussion is not to force a consensus but rather to ensure that all relevant information has been shared among the estimators and to reduce the chance of bias. After the discussion, estimators are offered the opportunity to change their estimates in light of new information. The changes are recorded and consensus estimates are formed. One method of forming consensus estimates is to suggest the median estimate of each quantile. Like a jury, if consensus is not reached, you have a "hung jury" and might need to declare a mistrial—to our knowledge, this has not happened.

In each case, the scientists must weigh the geoscience and exploration information. A number of the guidelines for making these estimates listed in table 8.2 were used by a team estimating undiscovered resources in Nevada (Cox et al., 1996). Some estimators used the number of known deposits per unit area of exposed permissive rocks multiplied by the area of permissive rock concealed by less that 1 km of postmineral rocks and sedimentary deposits, as in the above example of porphyry copper deposits. Some made their estimates on the basis of number of deposits known in well-studied areas of similar geology elsewhere in the world. Others depended on the number of occurrences that might become deposits as a result of more complete exploration, and still others were influenced by the number of exploration "plays" that could be visualized for the deposit

Table 8.2. Guidelines for number of deposits estimates.

Guideline	Example	References
Grade-and-tonnage model	All	
Frequency of deposits from well-explored areas (deposit density)	U.S. Alaska	MacKevett et al. (1978) etc.
	Western USA	Drew & others (1986)
	Costa Rico	Singer (1994a)
	Venezuela	Cox (1993)
	Australia	Scott (2000)
	General	Bliss & Menzie (1993)
		Singer & others (2001)
		Singer Berger, Menzie, and Berger (2005)
		Mosier & others (2007)
		Lisitsin & others (2007)
Local deposit extrapolations	U.S. Alaska	Singer and MacKevett (1977)
	U.S. Alaska	Root & others (1992)
	Japan	Kouda and Singer (1992)
Counting and assigning probabilities to anomalies	U.S. Alaska	Reed & others (1989)
	Puerto Rico	Cox (1993)
Process constraints	Worldwide	Drew and Menzie (1993)
Relative frequencies of related deposit types	Worldwide	Drew and Menzie (1993)
Area spatial limits	Worldwide	Singer and Mosier (1981b)
Ratio of occurrences to deposits	Worldwide	Anderson (1982)
Total known metal	U.S.A	
Statistical guides—coefficient of variation	All	Singer and Menzie (2005)
Exploration extent and efficiency	All	

type in question. Even where estimation guidelines and density of deposits models are available, it seems prudent to rely on mineral deposit specialists to make or modify the final estimates because they can bring their experiences and observations to the process. Evidence from experienced experts making independent estimates under these conditions indicates consistent estimates (figure 8.8).

Guidelines

Estimates are by deposit type and must be consistent with the grade-and-tonnage model and not with the population of mineral occurrences or weak

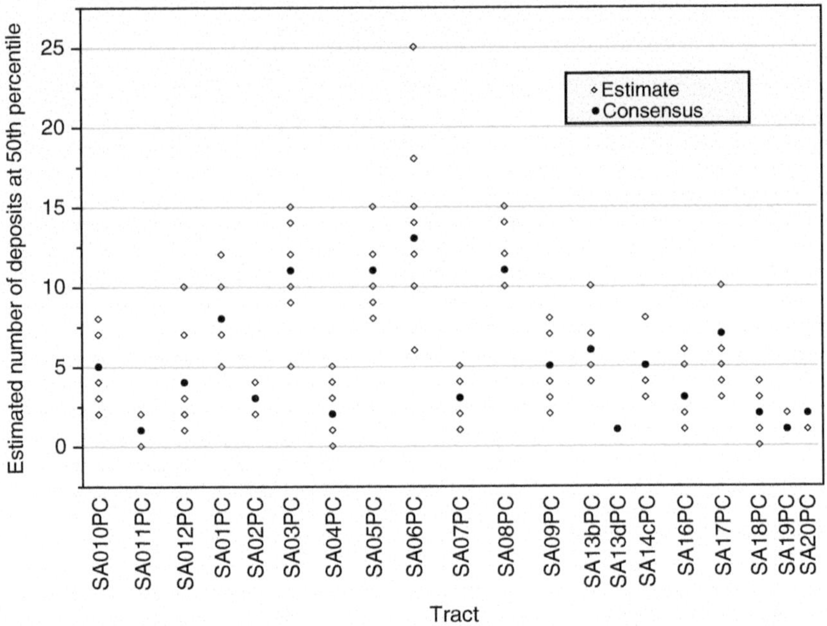

Figure 8.8 Comparison of 50th percentile estimates of the number of undiscovered porphyry copper deposits by each expert and the consensus estimates for tracts in South America. After Cunningham et al. (2008).

manifestations of an ore-forming process (Singer, 1994b). Thus, the estimated number of deposits must match the percentile values of the grade-and-tonnage model. For example, for any estimate, approximately half of the estimated undiscovered deposits should be larger than the median tonnage, and about 10 percent of the deposits should be as large as the upper 10 percent of the deposits in the tonnage model. If the grade-and-tonnage model is constructed with district data, then the number of undiscovered districts should be estimated. Some mineral deposit models, such as kuroko massive sulfide deposits, were constructed with spatial distance rules such as a 500-m rule for combining mineralization—the same rule must be applied when the number of undiscovered deposits is estimated. Well-explored deposits in the study area that have published grades and tonnages are counted as discovered deposits, whereas those without published estimates are counted as undiscovered in order to avoid double counting. The consistency of grade and tonnages to the number-of-deposits requirement is the most important guideline for making estimates. The expected number of deposits is not recommended for this comparison with the grade-and-tonnage distribution because it is the numbers of deposits at the 90 percent quantile, 50 percent quantile, and at the 10 or 1 percent quantiles that are estimated and adjusted if necessary.

There are no fixed methods for making estimates of number of undiscovered deposits. On the basis of experience and logic, however, a number of methods can be used directly or as guidelines to make these estimates (table 8.2). In most three-part assessments, the final estimates were made subjectively by experts and many have employed one or more of the methods in table 8.2 as guidelines.

Most guidelines represent some form of analogy. Most robust of these is a form of mineral deposit model wherein the number of deposits of each type per unit area from well-explored regions (Bliss and Menzie, 1993; Singer et al., 2001; Singer, Berger, Menzie, and Berger, 2005; Mosier, Singer, and Berger, 2007) is counted and the resulting frequency distribution is used either directly for an estimate or indirectly as a guideline in some other method (see chapter 4). Although Allais (1957) employed this method of estimating number of undiscovered deposits, many kinds of deposits were mixed together in his analysis.

Recognition that mineral deposits are uncommon and economic deposits are even less common was used by the Homestake Mining Company as a guideline in exploration planning because the success of an exploration program is dependent upon both the probability of discovering deposits and the value of the deposits that are discovered (Anderson, 1982). Homestake used published studies to estimate the probability of success at different stages of exploration. The largest study on metal exploration reported 5,718 reconnaissance examinations (approximately occurrences), 536 targets to be drilled, and 83 deposits developed (Anderson, 1982). Thus, we have a general guideline that between 1 and 10 percent of occurrences might be mineral deposits consistent with grade-and-tonnage models.

The coefficient of variation can be used to explore a variety of estimates between the extremes from either the deposit-density–based estimates or the Poisson distribution. A Poisson distribution with the same mean as estimated using deposit-density regression leads to estimates having a lower coefficient of variation and implies no clustering of deposits, whereas the regression approach suggests clustering of deposits. These statistical guides allow simple estimation of the number of undiscovered deposits in exposed or covered permissive terranes. This is accomplished by first estimating the standard deviation of the number of deposits using the following equation:

$$s_x = 0.121 - 0.237\,N_{90} - 0.093\,N_{50} + 0.183\,N_{10} + 0.073\,N_{05} + 0.123\,N_{01}, \quad (8.2)$$

where N_x = the estimated number of deposits at the *x*th percentile. If only the 90th, 50th, and 10th percentiles have been estimated, use the 10th estimate for the 5th and 1st also. A dimensionless measure of the spread of a distribution, the coefficient of variation, can be used to aid in selection of an appropriate distribution and to represent the uncertainty of the estimated number of deposits. It is defined as s_x/λ (λ is the expected number) and is represented here as percent relative variation.

AN EXAMPLE

In the state of Nevada there are seven known deposits that are defined in the same way as deposits in the porphyry copper descriptive and grade-and-tonnage models (see Singer et al., 2001). The tract permissive for all pluton-related deposits, including porphyry copper, covers about 41 percent of the area of the state (Cox et al., 1996). The well-explored, exposed permissive rocks in Nevada cover an area of about 32,800 km^2; five of the known porphyry copper deposits are in this exposed region. Areas covered by more than 1 km of material are excluded from consideration. Concealed permissive areas within 1 km of the surface are about 84,500 km^2 in extent. Two of the known porphyry copper deposits are completely covered by younger materials and cannot be considered to belong to the population of deposits that are well explored and exposed. If we assume that there are no additional porphyry copper deposits to be discovered in the exposed plutons in Nevada, then five deposits per 32,800 km^2 (the exposed permissive area) equals 0.00015 porphyry copper deposits/km^2. We can use this density of deposits to estimate the expected (mean) number of undiscovered porphyry copper deposits in Nevada. Thus, 0.00015 porphyry copper deposits/km^2 times 84,500 km^2 of covered permissive area equals an expected 12.9 concealed deposits, minus the two discovered deposits, which leaves eleven undiscovered concealed deposits that are defined in the same way as the deposits in the porphyry copper grade-and-tonnage model. Using the regression equation for porphyry copper deposits (equations 4.1, 4.2) leads to estimates of 6, 12, and 26 or more deposits at the 90, 50, and 10 percentiles. After subtracting the two known deposits, the estimates are 5, 10, and 22 or more deposits, with an expected number of deposits of eleven using equation 8.1. For comparison purposes, the subjective estimate of the expected number of undiscovered porphyry copper deposits in Nevada by Cox et al. (1996) is nine deposits. Thus, using a relatively local extrapolation of deposit density and a global porphyry copper density equation leads to expected number of undiscovered deposit estimates of eleven deposits, which is almost the same as the subjective expert estimate of nine deposits. These estimates are not completely independent because one of the experts used the local extrapolation to guide his independent estimates, but the local Nevada extrapolation was not used in development of the global equations (equations 4.1, 4.2).

AN EXAMPLE

In an unpublished study, four geoscientists made subjective probabilistic estimates of the number of undiscovered hot-spring mercury deposits in a 1:250,000-scale quadrangle in Alaska. They made independent estimates at the 90th, 50th, and 10th percentiles (table 8.3). The 10th percentile, for example, is the number of deposits for which there is at least a 10 percent chance of that number of deposits or more.

It was pointed out to participant D that because the number-of-deposit estimates must be consistent with the grade-and-tonnage model, his estimates imply that there is more undiscovered mercury in this quadrangle than has been found in the world in this deposit type. He replied that he was estimating wisps of cinnabar, not deposits consistent with the grade-and-tonnage model. In this case, knowledge of the total known amount of metal provided a guide to a flawed estimate by an economic geologist. Using a variety of different guidelines for estimates provides a useful cross-check of assumptions that may have been relied upon.

Table 8.3. Independent estimates by four scientists of the number of undiscovered hot-spring Hg deposits in a quadrangle in Alaska.

Participant	A	B	C	D	
90 percent chance of at least	1	1	2	9,000	deposits
50 percent chance of at least	3	2	4	10,000	deposits
10 percent chance of at least	6	6	7	11,000	deposits

AN EXAMPLE

In a study of undiscovered resources of Medford, Oregon (Singer et al., 1983), two geoscientists made subjective probabilistic estimates of the number of undiscovered kuroko massive sulfide deposits (table 8.4).

Estimates by expert B reflected the twenty-seven known kuroko-type occurrences in this 1:250,000-scale quadrangle. When expert B was asked whether half of each of his estimates would be larger than the median 0.3 million tons of the Sierran kuroko model, he said no and revised his estimates to a number similar to those of expert A. Having a common understanding of the need for consistency of the number-of-deposit estimates with the grade-and-tonnage model encouraged a consensus estimate to be made.

(*continued*)

AN EXAMPLE (*continued*)

Table 8.4. Preliminary independent estimates by two geologists of the number of undiscovered kuroko massive sulfide deposits in Medford, Oregon.

Participant	A	B	
90 percent chance of at least	1	27	deposits
50 percent chance of at least	2	50	deposits
10 percent chance of at least	5	110	deposits

AN EXAMPLE

In an unpublished study, one geologist estimated the number of undiscovered Climax porphyry molybdenum deposits in one 1:250,000-scale quadrangle of Colorado (table 8.5).

If the estimates were consistent with the grade-and-tonnage model, then the expected amount of molybdenum in deposits of this type in this single quadrangle would be two times the amount found in all known deposits. The estimates reflect the belief that every exposed and unexposed pluton might contain very low-grade molybdenum mineralization in the 1:250,000-scale quadrangle. Thus, the amount of molybdenum that is known to exist globally demonstrates that this set of estimates of number of undiscovered deposits is unrealistic, and the geologist should reexamine the grade-and-tonnage model and the implications of his estimates.

Table 8.5. Estimates by a geologist of the number of undiscovered Climax porphyry Mo deposits in Leadville, Colorado.

Participant	A	
90 percent chance of at least	4	deposits
50 percent chance of at least	20	deposits
10 percent chance of at least	38	deposits

AN EXAMPLE

In a study of tin resources of Alaska, the method of assigning probabilities to anomalies was used to estimate the number of undiscovered deposits (Reed et al., 1989). The Seward Peninsula in Alaska contains two known lode Sn deposits, a skarn (Lost River) and a stockwork/greisen (Kougarak), and a number of Sn occurrences that are associated with highly evolved 72- to 80-million-year-old epizonal biotite granites that intrude rocks of the York and Seward terranes. The York terrane consists of carbonate rocks of Ordovician to Mississippian age. The Seward terrane is primarily composed of early Paleozoic and possibly Precambrian schist and marble of blueschist, greenschist, and amphibolite facies. Evolved granites or their contact zones crop out at Cape Mountain, Potato Mountain, Ear Mountain, Lost River, Black Mountain, and Serpentine Hot Springs. Only the upper contact of the granite is exposed at most of these occurrences. This indicates that erosion is just now beginning to expose parts of the intrusive systems that could contain Sn deposits. It also suggests that other evolved granites may occur but not crop out. Industry had conducted a detailed aeromagnetic survey of the region that supplemented the USGS gravity survey.

An integrated geophysical model was used to delineate known granites and to identify concealed granites. Evolved granites were identified at ten locations (table 8.6). The geology of the Seward Peninsula is permissive for tin vein, greisen, skarn, and replacement deposits. A decision tree was used to evaluate the geologic, magnetic, and gravity data, and identify which types of deposits could occur around each of the granite plutons, and examine probabilities of deposits occurring.

Not all targets are necessarily deposits, and some deposits may not be identifiable by methods used. On the basis of the assigned target probabilities and number of targets, the authors made subjective estimates of 5, 9, and 15 or more skarn tin deposits at the 90, 50, and 10 percentiles.

Table 8.6. Estimation of number of skarn tin deposits in the Seward Peninsula, Alaska (after Reed and others, 1989).

Pluton	Number Targets	Probability	Expected Number
Cape Mtn.	2	1.0	2
Potato Mtn.	0	—	0
Ear Mtn.	2	1.0	2
Lost River	3	0.66	2
Black Mtn.	1	1.0	1
Divide	1	0.66	0.66
American River	1	0.5	0.5
Kougorak	2	0.5	1.0
Harris Dome	1	0.5	0.5
Serpentine Hot Springs	1	0.5	0.5
Total			10.2

AN EXAMPLE

In a 50,000 km^2 region permissive for porphyry copper deposits, the regression density model (equations 4.1, 4.2) would give 90, 50, and 10 percentile estimates of 5, 10, and 20 deposits, respectively, and an expected number of 10.2 deposits from equation 4.3. This regression distribution has a coefficient of variation of 82 percent. Using the expected number of deposits from the above regression, a Poisson distribution has estimates of 7, 11, and 15 deposits, respectively, and a much smaller coefficient of variation of 31 percent (table 8.7). Between these two end-member coefficients of variation there is a complete range of possible distributions that can be represented by either a negative binominal distribution or a distribution-free model generated by MARK3.

Only two of the many possible intermediate models are presented in table 8.7. Possible estimates using MARK3 are placed in equation 8.1 to ensure that they generate the same expected number of deposits as the density regression (equation 4.3), and acceptable estimates are placed in equation 8.2 so the coefficient of variation can be calculated. Possible estimates using the negative binominal distribution are generated by varying p and k parameters so that the expected value matches that of the density regression. With a coefficient of variation of 82 percent, the regression estimates can be viewed as representing the most uncertainty about the true number of deposits and that the deposits are clustered, whereas the Poisson distribution suggests less uncertainty and no clustering of deposits. Although the negative binomial distribution can be used for quantile estimates at intermediate coefficients of variation, it is somewhat awkward to work with, whereas MARK3 regression estimates from equations 8.1 and 8.2 are quite easy to generate.

Table 8.7. Some quantile estimates of the number of deposits and coefficients of variation of select distributions, where the expected number of deposits is 10.2, and $p = 0.163$, $k = 2$ for the negative binomial distribution.

Percentile	Regression	Negative Binomial	MARK3	Poisson
90th	5	3	5	7
50th	10	9	10	11
10th	20	23	16	15
5th	25	28	20	16
1st	39	39	27	19
Coefficient of variation (%)	82	77	55	31

AN EXAMPLE

In the example below, from Antarctica, no known porphyry copper deposits have been drilled because mining exploration is forbidden there (Singer et al., 2005). In Antarctica there is a belt of rocks similar to those in the Andes of South America. The widespread Mesozoic and Cenozoic intrusive and volcanic calc-alkaline rocks in the Antarctic Peninsula and eastern Ellsworth Land are considered a southward extension of the Andes of western South America (Rowley et al., 1983). On the basis of a 1:5,000,000-scale geologic map by Craddock (1972), the belt is about 1.7 million km^2 in extent, including shelf ice, permanent snow and ice, and water. The land areas, including areas mapped as permanent snow and unknown geology, cover about 1 million km^2. Exposed permissive parts of the Andean belt in Antarctica are about 76,000 km^2 in extent. Using the regression equations for porphyry copper deposits (equations 4.1, 4.2) leads to estimates of 90 percent, 50 percent, and 10 percent chance of at least 5, 12, and 24 undiscovered porphyry copper deposits, respectively, in this belt.

Although no porphyry copper deposits have been thoroughly explored in Antarctica, a number of prospects have been documented. Pride et al. (1990) examined four prospects in the northern Antarctic Peninsula, and three additional prospects are documented by Rowley et al. (1988, 1975). Rowley, Williams, and Pride (1991) summarize these prospects and discuss about twenty-five additional occurrences that have some characteristics in common with porphyry copper deposits. Thus, the estimates made with deposit densities of number of undiscovered porphyry copper deposits that are consistent with the grade-and-tonnage models certainly seem reasonable.

Suspicious Characters

Numerals representing estimated number of deposits provide information that can be helpful in refining estimates. The discussion in chapter 6 on significant digits in grade-and-tonnage models is relevant to the same issue with number-of-deposit estimates. Specifically, use of any more than two significant digits is misleading when reporting the expected number of deposits as estimated from equation 8.1 or the MARK3 program (chapter 9).

Much more subtle are the implied precision and knowledge expressed in the number-of-deposit estimates. When the numbers estimated at the 90th, 50th, and 10th percentiles are all greater than zero, the shape of the frequency distribution is fairly well determined, and the expected number of deposits cannot be greatly changed by estimates at the 5th and 1st percentiles. For example, estimates of 10, 15, and 20 deposits at the 90th, 50th, and 10th percentiles have fourteen expected deposits when the 5th and 1st percentile estimates are 20 and 20, but only fifteen expected deposits when the 5th and 1st estimates are 25 and 40 deposits. Furthermore, the estimates of 25 and 40 deposits imply a great deal more precision and knowledge than suggested by the estimates associated with the 90th, 50th, and 10th percentile estimates. Estimates at the 5th and 1st percentiles are intended for use when estimates at the 90th and 50th percentiles are zero so that the shape of the distribution can be simulated (chapter 9).

Risk Factor

If density models (chapter 4) are used to estimate the number of deposits, then a risk factor that expresses the probability that a region might not contain deposits might be necessary. The density models in chapter 4 were constructed in a way that excludes barren areas, so when a density estimate is made for a tract for which there is some chance of no deposits, that probability of zero deposits should be explicitly estimated.

Bias in Estimation

Bias exists when estimates consistently deviate from the true value (chapter 1). In estimating undiscovered mineral deposits, it is rare when the true number of undiscovered deposits is known in any useful time frame, so there is a question of how one might determine whether estimates are biased. As noted by Meyer and Booker (2001), estimation process issues can point to bias, such as the expert not following normative statistical or logical rules. These kinds of bias might be identified, for example, when the expert has a high estimate for one deposit type in a given tract and a low estimate for the same deposit type in another tract that is stated by the expert to be not different. Another example would be when the expert states that the deposits are clustered in space and yet makes number-of-deposit estimates that are consistent with a Poisson distribution, suggesting spatially independent deposit locations.

The most straightforward way to identify bias in estimates of number of deposits is to use the same tools that are available for guidelines for the estimates. Number-of-deposit estimates that are inconsistent with the guidelines should at the least raise questions about possible bias. If the estimates are noticeably different than one or more guidelines, one should definitely reexamine the estimates and/or the guidelines. In our experience, the problem is typically due to an error in an underlying assumption of the assessor. Typical reasons for overestimation of undiscovered deposits include (1) estimating the number of occurrences or geochemical anomalies instead of mineral deposits, (2) estimating the number of occurrences or "deposits" that have tonnages greater than some minimum, and (3) political considerations. The estimated undiscovered deposits must be considered to be from the same frequency distributions as the grade-and-tonnage model in order for the estimate of resources to be unbiased.

Using a minimum tonnage to define a deposit (either known or undiscovered) can cause a biased estimate of the number of undiscovered deposits because a large number of incompletely explored mineralized systems may be known that can never fit the grade-and-tonnage model of mineral deposits. That is, some of the estimated "deposits" will be from the upper tail of the frequency of incompletely explored mineralized systems and not from the same frequency distribution of tonnages of true mineral deposits. Therefore, resources calculated from the combined number-of-deposit estimates and the grade-and-tonnage model will be biased upward.

Lack of consistency between estimates of number of undiscovered mineral deposits and grade-and-tonnage models can lead to bias in the form of severe overestimation of undiscovered resources. To insure unbiased resource estimates, the estimated undiscovered deposits must have the same frequencies of tonnages and grades as the grade-and-tonnage model. This means not only that 90 percent of the estimated deposits are as large as the largest 90 percent of deposits in the model, but 50 percent are as large as the median model tonnage, and 10 percent are as large as the largest 10 percent in the model. This relationship holds for each estimate of number of undiscovered deposits regardless of the certainty of the estimate. In addition, some models were constructed with spatial distance rules such as the 500-m rule for combining mineralization in the kuroko massive sulfide model (Mosier, Singer, and Salem, 1983)—these same rules must be applied to the number of undiscovered estimates. Use of the guidelines is crucial to the generation of unbiased estimates and to the identification of possible biases.

For example, by 1996, about 5,500 metric tons of gold had been discovered in various kinds of deposits in Nevada. In an assessment of undiscovered base and precious metal-bearing deposits in that state, Cox et al. (1996) and Singer (1996), estimated about 4,000 tons of gold in deposits in the upper 1 km. This total estimate can be derived by summing the expected number of gold-bearing deposits times the expected average grades and the expected tonnage of each deposit type. According to an economic geologist critical of this form of assessment and knowledgeable about Nevada, these estimates of undiscovered gold are low by a factor of 100. Thus, he estimates that there are 400,000 tons of gold yet to be discovered in Nevada. If he were right, the 400,000 tons of gold in Nevada would represent about two times the total amount of gold discovered through history in the whole world (Singer, 1995). In this case, the guideline of global content is used to demonstrate that a subjective estimate made without critical thinking can be biased.

Some estimators assume that it is better to make conservative or liberal estimates to avoid errors. Conservative or liberal in this context are just other words for biased, which is a serious error. Logic dictates that estimates be unbiased and explicitly state the uncertainty. Fortunately, number-of-deposit estimates in three-part assessments are very hard to bias because they must be consistent with the grade-and-tonnage model, the descriptive model, and all information available, and they are typically made by consensus.

Summary

The goal of many assessors is to make unbiased quantitative assessments in a format needed in decision-support systems so that the consequences of alternative courses of action can be examined. Internally consistent descriptive, grade-and-tonnage, deposit-density, and economic models, and the design of three-part assessments reduces the chances of biased estimates of the undiscovered resources. Biases can be introduced into these estimates either by a flawed grade-and-tonnage model or by the lack of consistency of the grade-and-tonnage model with the number-of-deposit estimates.

Estimates of number of undiscovered deposits explicitly represent the probability (or degree of belief) that some fixed but unknown number of undiscovered deposits exist in the delineated tracts. As such, these estimates reflect both the uncertainty of what may exist and a measure of the favorability of the existence of the deposit type. Although there are no fixed methods for making estimates of number of undiscovered deposits, guidelines to help make these estimates are available. Using a variety of different guidelines for estimates both provides a useful cross-check of assumptions that may have been relied upon and significantly reduces the chances of biased estimates. Guides include the following:

1. The deposit frequency distribution is expected to be either Poisson or, perhaps more likely, negative binominal with heavy right tails.
2. About half of each estimate should be larger than the median tonnage and median grade of the deposit type being estimated.
3. Rough estimates can be made with the universal regression equation that relates permissive area and deposit size to number of deposits.
4. More refined estimates are available as deposit-density models for some deposit types.

Until new more refined estimation guidelines are available, it seems prudent to rely on the universal regression equation followed by mineral deposit specialists to make subjective estimates because the specialists can apply their experiences and observations to the process. This kind of activity is not unusual; geologists commonly make estimates that, although not explicitly quantitative, are subjective and have uncertainty, such as making geologic cross sections. Experience from meteorology suggests that consensus schemes perform better than most individual estimators, and the best estimates are made when objective estimates such as those from guides (table 8.2) are part of the information supplied to subjective estimators.

Sensitivity analysis of porphyry copper deposits shows that the greatest opportunity for reducing uncertainty in exploration and resource assessment lies with lowering the uncertainty associated with tonnage estimates. This means that selection of the proper grade-and-tonnage model is probably more critical to the final assessment than small errors in the number-of-deposit estimates. By following the procedures presented here and consistently applying the guidelines, significant biases in estimated number of undiscovered deposits are unlikely.

Once estimates of the number of undiscovered deposits have been obtained, the quantification of undiscovered mineral resources using the three-part methodology is complete. Depending upon what question the assessment is to answer, additional analysis might be required. It is very likely that data on discovered resources will need to be gathered to complete the picture of the mineral resources of the region. Also it may be necessary to combine the estimates of undiscovered deposit with the grade-and-tonnage data via Monte Carlo simulation and to evaluate the proportion of the resulting resources that will be economic under assumed conditions. We have already discussed the use of engineering cost models (chapter 5) to perform part of this analysis, and refer the reader to Root, Menzie, and Scott (1992) for a detailed discussion of simulation and chapter 9 for an overview.

9

Integration of Grades, Tonnages, Number of Deposits, and Economic Effects

Perspective

Now that all of the fundamental parts of a quantitative mineral resource assessment have been discussed, it is useful to reflect on why all of the work has been done. As mentioned in chapter 1, it is quite easy to generate an assessment of the "potential" for undiscovered mineral resources. Aside from the question of what, if anything, "potential" means, there is the more serious question of whether a decision-maker has any use for it. The three-part form of assessment is part of a system designed to respond to the needs of decision-makers. Although many challenging ideas are presented in this book, it has a different purpose than most academic reports. This book has the same goal as Allais (1957)—to provide information useful to decision-makers. Unfortunately, handing a decision-maker a map with some tracts outlined and frequency distributions of some tonnages and grades along with estimates of the number of deposits that might exist along with their associated probabilities is not really being helpful—these need to be converted to a language understandable to others. This chapter summarizes how these various estimates can be combined and put in more useful forms.

If assessments were conducted only to estimate amounts of undiscovered metals, we would need contained metal models and estimates of the number of undiscovered deposits. Grades are simply the ratio of contained metal to tons of ore (chapter 6), so contained metal estimates are available for each deposit (table 9.1).

Table 9.1. Grades, tonnages, and contained copper in hypothetical grade-and-tonnage model.

Deposit	Tons	Cu Grade %	Tons Cu
A	6,400,000	1.7	108,800
B	221,000	1.2	2,652
C	93,000	4.37	4,064
D	1,260,000	1.6	20,160
E	447,000	0.71	108,800
		Mean =	27,770

In the simplest of all cases, one could estimate the expected number of deposits with equation 8.1 (see chapter 8) and multiply it by the expected amount of metal per deposit, such as the 27,770 tons of copper in table 9.1, to make an estimate of the expected amount of undiscovered metal. As pointed out in chapter 1, expected amounts of resources or their values can be very misleading because they provide no information about how uncommon the expected value can be with skewed frequency distributions that are common in mineral resources; that is, uncertainty is ignored. In the next simplest case, one could use a Poisson distribution to represent the number of undiscovered deposits with a mean equal to the expected number of deposits. A lognormal distribution with the same mean and standard deviation as the logged metal content in the model deposits would be used to represent contained metal in undiscovered deposits. This situation can be solved analytically as shown by Allais's study of the Algerian Sahara (Allais, 1957), where he used value rather than contained metal. This formulation could also be solved with a simulator that would take little time to prepare and run because only two standard distributions are used and independence between them is reasonable to assume. It is tempting to think that one could just add or, if appropriate, multiply the percentiles of the various frequency distributions together. Unfortunately, this is very dangerous mathematically because in most situations these percentile values are not additive, as shown in the example of figure 9.1. Simulation is required in these situations.

In the next level of complexity, a Poisson distribution would again represent the number of undiscovered deposits and lognormal distributions would represent the distributions of tonnage of ore and of grades with the assumption of independence of grade and tonnage (see chapter 6 for details). This can also be solved analytically for metal content or gross value but may be misleading if there are any significant correlations among the variables. Taking correlations into account increases the complexity to the point where simulation is preferred. This is the formulation used by Drew et al. (1986) in an early version of a Monte Carlo simulator called MARK3. This method also underestimates total uncertainty if the experts believe that there is

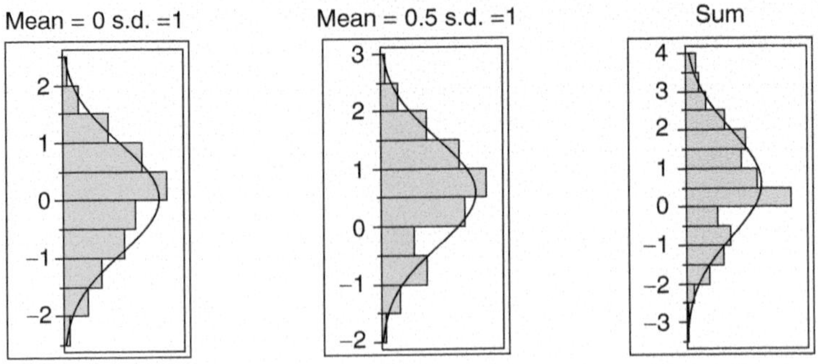

Quantiles		Quantiles		Quantiles		SUM
90%	1.14	90%	1.76	90%	**2.43**	**2.90**
50%	0.086	50%	0.632	50%	**0.555**	**0.718**
10%	−1.34	10%	−0.756	10%	**−1.25**	**−2.09**
Mean =	0.022	Mean =	0.579	Mean =	**0.601**	Mean = **0.601**
N =	200	N =	200	N =	200	

Figure 9.1 Simulations of two random normal distributions (mean = 0.0, standard deviation = 1, mean = 0.5, standard deviation = 1) showing that although the means can be mathematically added (0.601 = 0.601), adding the 90th quantiles produces different results than obtained from the quantiles of the sum of the two variables (2.43 ≠ 2.90), as does adding the median (0.555 ≠ 0.718) and the 10th quantiles (−1.25 ≠ −2.09).

clustering of deposits because the Poisson distribution requires an assumption of independence of values in space, and it has a lower variance than do clustering distributions (see chapter 8).

Determining whether the undiscovered metals might be economic to recover is an important output of most assessments, and grades and tonnages are necessary inputs needed to estimate economic viability of mineral deposits (chapter 5). Thus, in order to be able to provide decision-makers with information about undiscovered mineral resources in a form demonstrating possible consequences of their decisions, it is necessary to have a general simulator that incorporates economic filters. How the economic effects are incorporated in a simulator is discussed after the simulation of tonnages and grades and number of undiscovered deposits are presented.

Number of Deposits

In chapter 8 we noted that estimates of number of deposits explicitly represent the probability (or degree of belief) that some fixed but unknown number of undiscovered deposits exist in the delineated tracts. Uncertainty is reflected by the spread of the number-of-deposit estimates (quantiles) associated with the 90 percent quantile to the 10 or 1 percent quantile, with a large difference in the numbers suggesting great uncertainty. To represent the probabilities of each possible number of deposits that may exist, we need to either have made individual estimates of the probability for each of these possible numbers of deposits or have a shortcut. For small numbers of deposits, estimating the probabilities might be reasonable, but for larger numbers of deposits, this procedure becomes impractical. If we were willing to make some assumptions, we might use the well-known Poisson distribution to estimate these individual probabilities. Because the assumptions required with the Poisson distribution are not always appropriate (see chapter 8), we use another, less restrictive way to estimate the probabilities for each possible number of deposits.

In the MARK3 program (Root, Menzie, and Scott, 1992), an algorithm was presented for estimating the probabilities associated with the intermediate numbers of deposits between estimated the number of deposits that may be present in the tract at a 90 percent level, a 50 percent level, and a 10 percent level. Interpretation of these quantile estimates requires care. Because of their "at least" format (Root, Menzie, and Scott, 1992), a unique underlying probability distribution cannot be specified on the basis of the elicited distribution alone. Root, Menzie, and Scott (1992) provide a methodology for obtaining a probability distribution that is consistent with these quantile estimates. The shaded areas in figure 9.2 indicate regions where distributions are consistent with 90, 50, and 10 percentile estimates of 1, 2, or 4 deposits. An infinite number of distributions are consistent with the three estimates of number of deposits. Root et al.'s method chooses a distribution that is approximately in the middle of all possible choices, as indicated by the MARK3 estimates at each number of deposits in figure 9.2.

In some cases, it is desirable to make estimates at the 5 and 1 percent levels, so here we include a version of Root et al.'s algorithm that has been expanded to include estimating probabilities associated with number of deposits estimated at the 90, 50, 10, 5, and 1 percent levels (see appendix 2). The five number-of-deposit estimates associated with these percent levels divide the possible nonnegative integers into six intervals: 0 to N(9), N(9) to N(5), N(5) to N(1), N(1) to N(05), N(05) to N(01), and N(01) to infinity. Integers in these six intervals receive 10%, 40%, 40%, 5%, 4%, and 1%, respectively of the unit probability. In Root et al.'s method of allocating probabilities, the numbers associated with N(9) and N(5) lie half in each of the two intervals of which they are endpoints, and they receive probability from each, but half of what an interior point gets. N(01) receives half of what the interval N(05) to N(01) receives plus 0.01. The largest possible number of deposits that is given a probability is that associated with N(01). An algorithm to allocate the total probability among all possible number of deposits is described in appendix 2.

An example of how the algorithm allocates the probabilities is shown in figure 9.3. Cumulative probabilities like those shown in figure 9.3 are

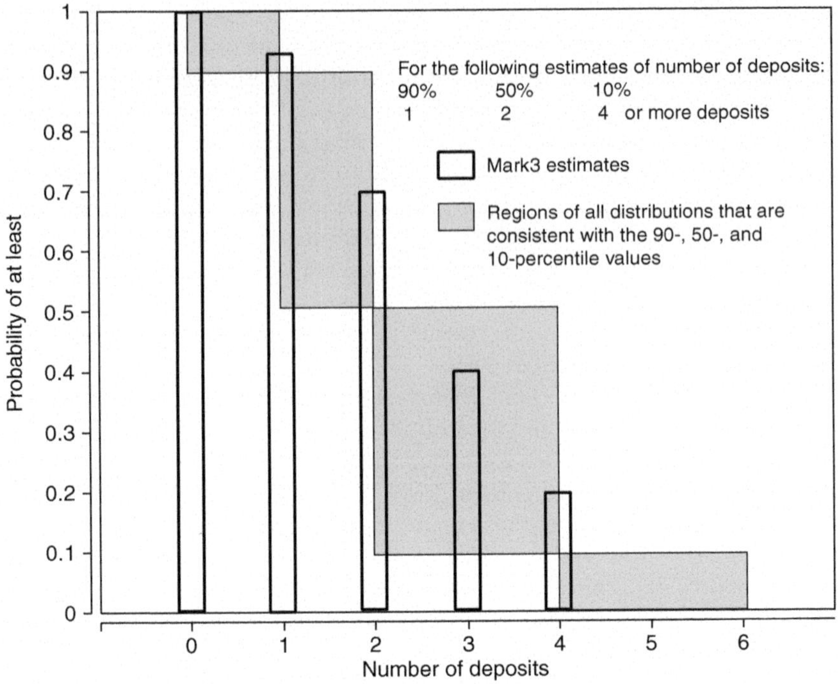

Figure 9.2 Regions of all possible distributions consistent with number-of-deposit estimates of a 90% chance of 1, a 50% chance of 2, and a 10% chance of 4 or more deposits and the MARK3 estimates. After Root, Menzie, and Scott (1992).

used in the MARK3 simulator to properly weigh the frequencies of different numbers of deposits, as illustrated in figure 9.4. There is an option in the simulator where the user can specify the probability of zero deposits: the probabilities for integers greater than zero are multiplied by a constant to keep the total probabilities equal to 1. In operation, the simulator starts by selecting a random number between 0 and 1 that is drawn from a uniform distribution. This random number is used to select a number of deposits from the y-axis shown in figure 9.4. The simulator then draws that number of tonnages from the distribution of tonnages in grade-and-tonnage model of the appropriate deposit type.

Tonnages and Grades

In each version of the MARK3 program, there are files representing the grades and tonnages of each deposit type that are the source data for the grade-and-tonnage models (chapter 6). The deposit data in these files are grouped

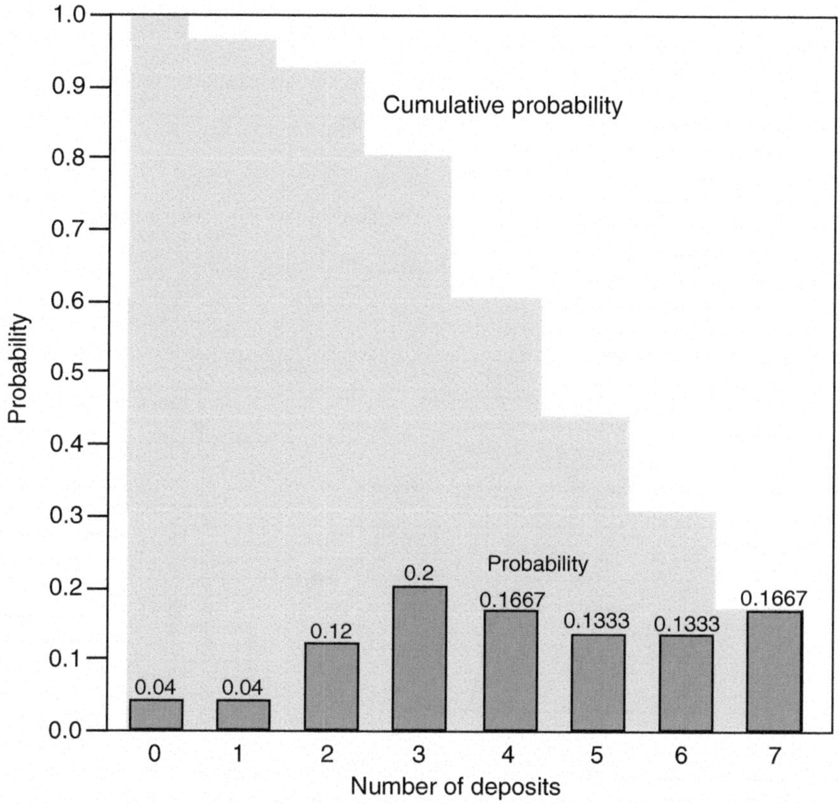

Figure 9.3 Probabilities and cumulative probabilities for number of deposits from MARK3 algorithm, where the estimates are 2 or more deposits at the 90% level, 4 or more at the 50% level, and 7 or more at the 10% level.

according to suites of metals that have been reported. For example, in table 9.2 the grade-and-tonnage data for 31 carbonatite deposits are grouped into a suite of 7 deposits with niobium and rare-earth grades reported, 17 deposits with only niobium reported, and 7 deposits with only rare-earth grades reported. Regardless of the kind of simulation performed in MARK3, these groups are selected in proportion to the frequency of deposits in each group. For instance, when a simulation requires the selection of a tonnage from a carbonatite deposit, 17 out of 31 times it will select from the group of deposit tonnages that only have niobium reported.

Using Root et al.'s methods, these suites form the basis for distributions and dependencies in two ways to perform the simulations of grades and tonnages. One option uses approximations of lognormal distributions, and the other uses approximating piecewise linear distributions. The second, more commonly used, is called the empirical option, and the first is called the lognormal option. The user of the simulation program selects the desired

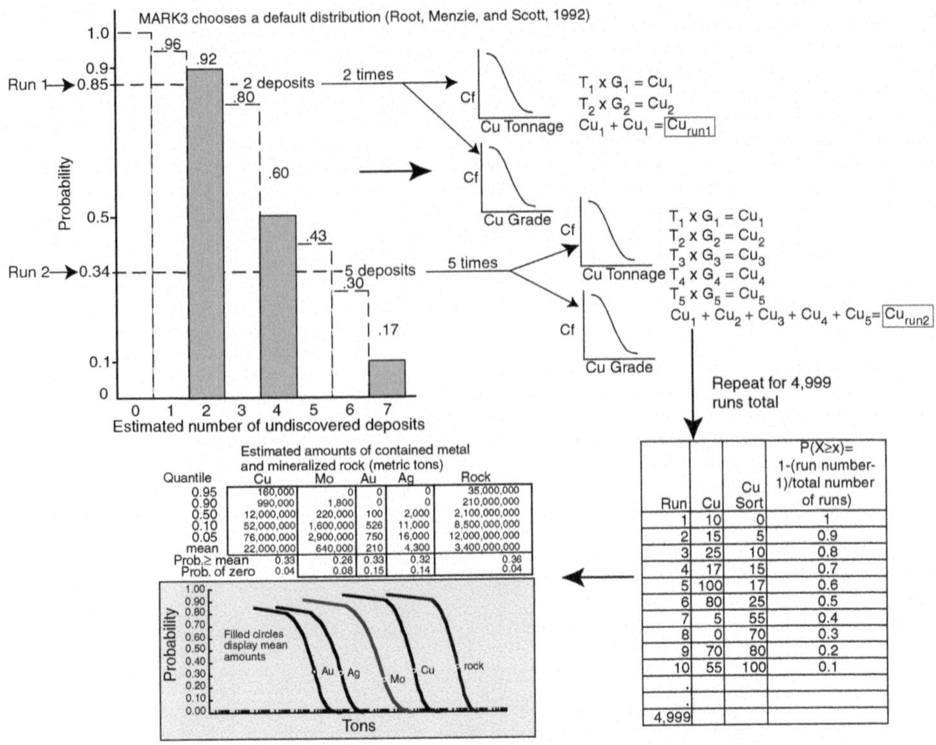

Figure 9.4 Schematic illustration of the MARK3 simulator when the estimated number of deposits is 2 at the 90% level, 4 at the 50% level, and 7 at the 10% level. The deposit type is porphyry copper. Cf stands for cumulative frequency. Original figure by W. David Menzie, Don Singer and Joe Briskey.

option. Because of the complexity of the algorithms needed to account for dependencies among the variables and for producing unbiased representations of the model grades and tonnages, a detailed description of the methods is not produced here, and the reader in referred to the original source (Root, Menzie, and Scott, 1992).

Regardless of which option is selected, the basic nature of the simulation can be illustrated in figure 9.4. As noted above, the Monte Carlo simulator draws random numbers between 0 and 1.0. For each of the 4,999 draws, a number of deposits is selected, and for that number of times, the simulator selects the suite and a tonnage and as many commodity grades as exist in the suite. Output from the simulator typically displays tables of frequencies of number of deposits, tonnages, and contained metals. In addition, plots of cumulative frequencies as shown in figure 9.4 are available. The procedures used in Root et al.'s algorithms properly represent the frequencies of number of tonnages, grades, and contained metals, as demonstrated by figure 9.5.

Table 9.2. Grade and tonnage data for carbonatite deposits grouped according to suites of metals reported.

Deposit	Metric Tons/10^6	Nb_2O_5 Grade %	RE_2O_5 Grade %
Bayan Obo	750	0.1	4.1
Araxa	462	2.48	0.033
Oka	221	0.23	0.1
Mrima Hill	50.8	0.67	0.59
Iron Hill	36.3	0.25	0.01
Catalao I	21	0.98	1.016
St. Honore (Soquem)	16	0.69	0.0088
Panda Hill	272	0.3	
Salitre II	200	2	
Tapira	166	1.18	
Martison Lake	140	0.35	
Sukula	118	0.25	
Mbeya	116.9	0.31	
Nemogos (Lackner Lake)	112.2	0.236	
Serra Negra	60	1.5	
Sove	55.3	0.23	
James Bay	36.3	0.52	
Dominion Gulf	33	0.39	
Lueshe	30	0.35	
Ondurukurme	7.2	0.3	
Bingo	7.1	2.86	
Catalao II	2	2.18	
Manitou Island	1.9	0.86	
Kaiserstuhl	0.63	0.5	
Phaloborwa	2210		0.57
Amma Dongar	105		3
Mountain Pass	90		5
Mushgia Khudag	6.1		1.37
Pococ de Cardas	6		5
Ludiin Gol	0.37		3.2
Kangankunde	0.3		4.7

Here, the average content of metals, calculated in the same way as in table 9.1, is plotted against the mean content of the same metals for several different deposit types; the correlation coefficient between the input metals and the simulated output metals is greater than 0.99, and the slope of the regression line is 1.0, demonstrating that the simulator introduces no bias. These simulations were done with the empirical option of the program (Duval, 2004). In general, the empirical option is most commonly selected largely because it so faithfully reproduces the original data. Software implementing these features is available in Bawiec and Spanski (2009).

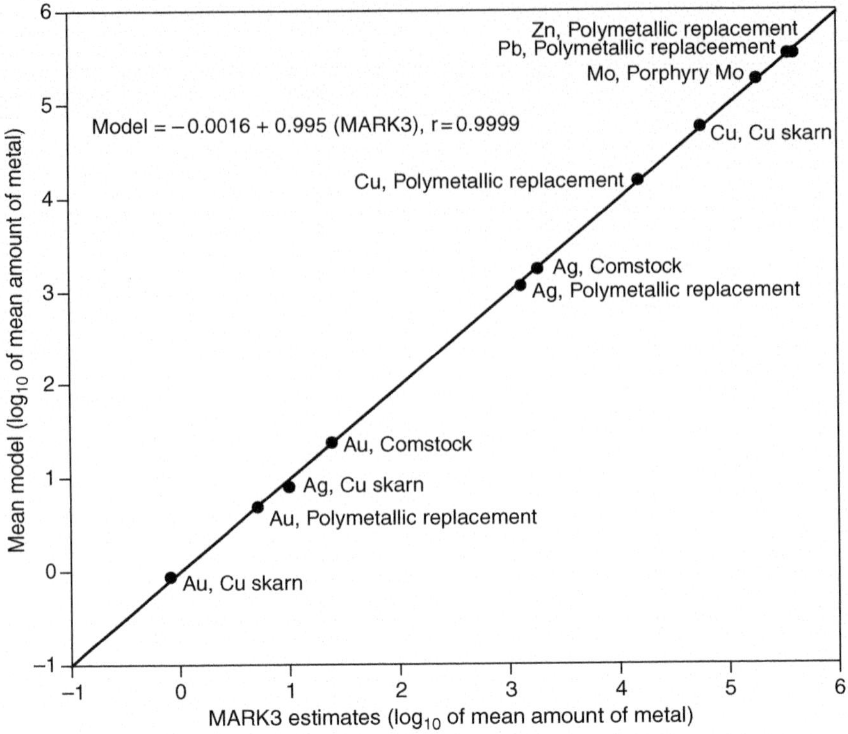

Figure 9.5 Test of possible bias in MARK3 simulations (2003 version).

ASSESSMENT EXAMPLE

In 1976, the U.S. Congress instructed the Bureau of Land Management (BLM), in the U.S. Department of Interior (DOI), to review public lands under its jurisdiction for suitability as wilderness areas. The U.S. Geological Survey (USGS) and the former U.S. Bureau of Mines (USBM), sister agencies of BLM in DOI, were required to perform mineral surveys on proposed lands to inform DOI, which would make recommendations about the lands to Congress. Reports from USGS and USBM on 2 million acres of BLM proposed wilderness areas were complete by the 1980s (Beikman et al., 1983). Areas were assessed as having high, medium, or low potential for mineral commodities or for types of mineral deposits. By 1985, officials in DOI began to question the utility of these qualitative assessments as a basis for making decisions on lands containing multiple resources.

Designating areas wilderness required developing an environmental impact statement (EIS) and holding public hearings. If USGS and USBM mineral surveys were completed before the final EIS,

BLM included these studies as part of the EIS. Many mineral surveys were completed after the EISs (Marcus et al., 1986), but surveys in Nevada and Colorado were completed in time in early 1990. In 1990, the Colorado BLM office finalized its recommendations without making changes based on the mineral surveys. The Nevada BLM office recommended changing four areas from suitable to unsuitable for wilderness, and changed boundaries of three other areas.

In May 1991, DOI instructed the USGS to perform quantitative assessments for nine areas in Nevada and Colorado previously determined to contain significant high or moderate potential for undiscovered or discovered mineral resources. Assessment teams performed three-part assessments only on deposit types identified in the original wilderness reports. Estimates of numbers of deposits were combined with appropriate grades and tonnages models using the MARK3 simulation program. Because the USBM did not have personnel resources to conduct economic evaluations using the potential supply methodology, gross-in-place values of undiscovered resources in the areas were calculated. In late June, results for the nine areas were transmitted to BLM and USBM and published in August after additional study areas in Oregon and Utah were evaluated at DOI's request (McCammon et al., 1991). The quantitative assessment report clearly stated that in four of the areas, there was a significant chance that the gross-in-place value of undiscovered mineral resources was negligible.

BLM took the position that changes to wilderness proposals formulated by their state offices might require a supplemental EIS and a period of public discussion. A Colorado newspaper, the *High Country News* (15 July 1991, p. 4), published an article where they quoted BLM officials opposing changes to the wilderness proposals and quoted representatives of environmental groups who characterized the review as a reevaluation, and expressed the opinion that it violated the law because it took place after the final EIS. On 3 September, an Assistant DOI Secretary determined that two Colorado, three Nevada, and six Utah wilderness study areas be excluded from DOI's wilderness proposals. BLM appealed the decision to the secretary of DOI, who agreed to reinstate one Nevada and five Utah wilderness study areas. In October 1991, several environmental groups filed suit against the secretary of DOI, asking that the secretary be found in violation of the law and quantitative assessments be released as supplemental EISs.

These policy and political actions were also joined with considerable criticism of quantitative mineral resource assessment following the decisions (see Drew, 1997). During the fall of 1991, criticism of the three-part quantitative assessment from within and outside the USGS was intense. In the face of this criticism, the USGS let a contract to conduct a review of the quantitative-mineral-resource assessment methodology. The review committee in its report (Harris and Rieber, 1993) concluded, "Many of the recent criticisms of the methodology are either fundamentally incorrect or exaggerated in terms of their technical

(*continued*)

ASSESSMENT EXAMPLE (*continued*)

merit." The panel recommended gross-in-place value be replaced by reduced-gross-in-place value for deposits that would be economic when potential supply analysis could not be made. National elections in 1992 led a new Secretary of the Interior, who overturned his predecessor's decision regarding the Colorado, Nevada, and Utah wilderness areas.

Economic Effects

Each of the 4,999 realizations of the simulation that produces a deposit (not all necessarily do) has a tonnage and one or more associated grades. Because only some of the deposits used to create grade-and-tonnage models were economic to mine, it is not clear which, if any, of the simulated undiscovered deposits might be economic to mine in the future. Simulated grades and tonnages along with assumed prices of metals allow the calculation of value per ton and gross value of each deposit. Costs of each of these simulated deposits could be estimated by using the cost models discussed in chapter 5.

Selection of the appropriate cost model depends on the nature of the deposit type being simulated and on location effects such as access and depth to the deposit. Each deposit type tends to have a particular shape and form, which are important determinants of appropriate mining methods. In addition, there is a history of mining methods linked to certain deposit types. Deposits that are remote and that do not have existing infrastructure are costlier to develop and mine, so these costs should be accounted for in some way. Depth of the deposit affects decisions about whether underground mining is required and costs associated with deeper mining, such as capital and operating costs. Because modern cost coefficients are still being researched, in 2008 there still was no publicly available version of software implementing economic filters.

10

Exploration Risk

Perspective

It is commonly said that mineral exploration is a risky business, but what does that really mean? Although exploration can be financially rewarding, there is a high probability that a single venture will be a failure. Risk is defined as chance of failure or loss and its adverse consequence (i.e., failure or loss). Risk differs from uncertainty in that uncertainty simply means lack of knowledge of the outcome or result, whereas risk involves a loss. Thus, one could be uncertain of an outcome, but not necessarily be at risk of losing something. In risk analysis, two quantities are estimated: the magnitude (severity) of the possible adverse consequence(s), and the likelihood (probability) of occurrence of each consequence. Procedures of risk analysis are well established, if not simple, and are applied in both business and engineering (Aven, 2003; Bárdossy and Fodor, 2004; Davis and Samis, 2006). Mineral exploration is an economic activity involving risk and uncertainty, so risk also must be defined in an economic context in which the extent of the loss is defined. Successful mineral exploration strategy requires identification of some of the risk sources and consideration of them in the decision-making process so that controllable risk can be reduced.

It is not uncommon to see recommendations that exploration firms should accept all projects with positive expected monetary values—that is, projects that have a positive economic value after being multiplied by the probability of deposit discovery and subtraction of exploration costs. Clearly, this strategy would be unwise for a firm with limited resources if the chance of failure were significant. Both expected monetary values and the probabilities of various outcomes such as economic failure should be considered in the decision-making process. Because economic return, when measured by net present value, is closely related to the size of mineral deposits, and because deposit sizes can be represented by highly skewed frequency distributions, achieving expected monetary or higher values tends to be a low-probability outcome. This and the typical rareness of mineral deposits are the fundamental reasons for the high risk in mineral exploration. Mineral exploration can be viewed as a process in which resources are successively expended to reduce the uncertainty about where, how large, and how rich undiscovered mineral resources are. It can be characterized as a multistage search process in which only the last stage, drilling or digging, is usually definitive. At each stage, an attempt is made to reduce the area to which the next, typically more expensive, stage of search is applied. Each stage may be viewed as an attempt to discriminate between areas that contain valuable deposits and areas that do not. Because deposit detection is probabilistic, at each stage, classification errors of both types (i.e., rejecting valuable deposits and accepting nonvaluable prospects) and their associated costs must be balanced against the possible gains of discovering an economic deposit.

The exploration department of a major zinc producer found it essential to document a robust decision-making process in order to maintain internal and investor support (Penney et al., 2004). Zinc deposits from around the world were classed by type, grade, and tonnage models developed for each, cost filters were applied to each, and tracts around the world were delineated where the types could occur (Penney et al., 2004). Their process was the same as that recommended in this book, with the exception that they ranked or scored tracts rather than estimating the number of undiscovered deposits.

There is a close connection between planning for exploration and the kind of mineral resource assessments discussed in this book in that the uncertainties are explicitly identified and quantified in these assessments. Exploration risk is a function of the rareness of mineral deposits and of the skewed distributions of tonnage and grade (chapters 2 and 6). Very large deposits are needed to sustain supply and are more likely than small deposits to have high net present values, and economic trends require larger sized deposits in order to take advantage of economies of scale (chapter 5). Many major exploration companies today require a minimum size of what is commonly called a "world-class" deposit. Exploration risks for these deposits can be reduced using strategies based on geology, statistics, and economics. The analysis by Penney et al. (2004) demonstrated the importance of including infrastructure costs in considering the economics in remote locations (see chapter 5). Selection of deposit types that had a high probability of being economic given seven infrastructure categories was a critical step in their decision-making about where to explore. Remote locations with no existing infrastructure significantly reduced the number of desirable deposit types.

Here we present the basic ideas of risk as they apply to exploration—a basic equation provides a way to examine risk in this chapter. The most direct way of reducing risk by increasing the number of prospects examined is considered first. Another way to reduce risk is lowering the probability of failure by directing exploration to deposit types that have a higher probability of being found. Risk reduction is then considered through using targeted geologic settings. Various means are also available to reduce economic risk in the exploration, development, and mining phases (Moore and Drew, 1979). Finally, we discuss risk reduction by using prior information and by changing the assumption of independence in a sequence of exploration examinations, that is, by learning. Although it is implicitly assumed that multiple decisions are made in an exploration program, some of the tools presented here can be used in small companies for single-event decisions. The ingredients for exploration risk analysis are the three parts of assessments that form the core of this book, along with the deposit models.

Basics of Risk

In its most elemental form, risk can be considered the probability of failure (and its associated cost). For example, if we predict heads when a fair coin is flipped, the risk of failure is that tails might be the result. In the long term, under these conditions, we expect that our probability of losing on any single coin toss will be 0.5. We can also ask what is the probability of two losses (tails) in a row. If the results of one coin toss have no effect on the next coin toss, the events are independent, and the probability of two losses is 0.25 (i.e., 0.5 × 0.5). We can also calculate the probability of at least one success (head) as 1 minus the probability of both coin tosses being tails (1.0 – 0.5 × 0.5), or 0.75. The same rules apply if there are three coin tosses; the probability of at least one success is 1.0 – 0.5 × 0.5 × 0.5, or 0.875. Generalizing this: the probability of at least one success can be estimated as 1.0 minus the probability of failure multiplied by itself as many times as there are trials:

$$P_{success} = 1 - (p_{failure})^n \quad (10.1)$$

where $p_{failure}$ is the probability of failure in one trial, and n is the number of trials. These simple concepts form the foundation for examining exploration risk and reducing it to acceptable levels. In the following sections, each part of this equation is examined.

Slichter (1960) said that the only way to avoid *gambler's ruin* in mineral exploration (equation 10.1) was to have enough capital to have many trials, that is, have a large n (see figure 10.1). Thus, the classic way to reduce risk in exploration is to increase the number of prospects examined (n). Consideration of this was central to the successful exploration and discovery of

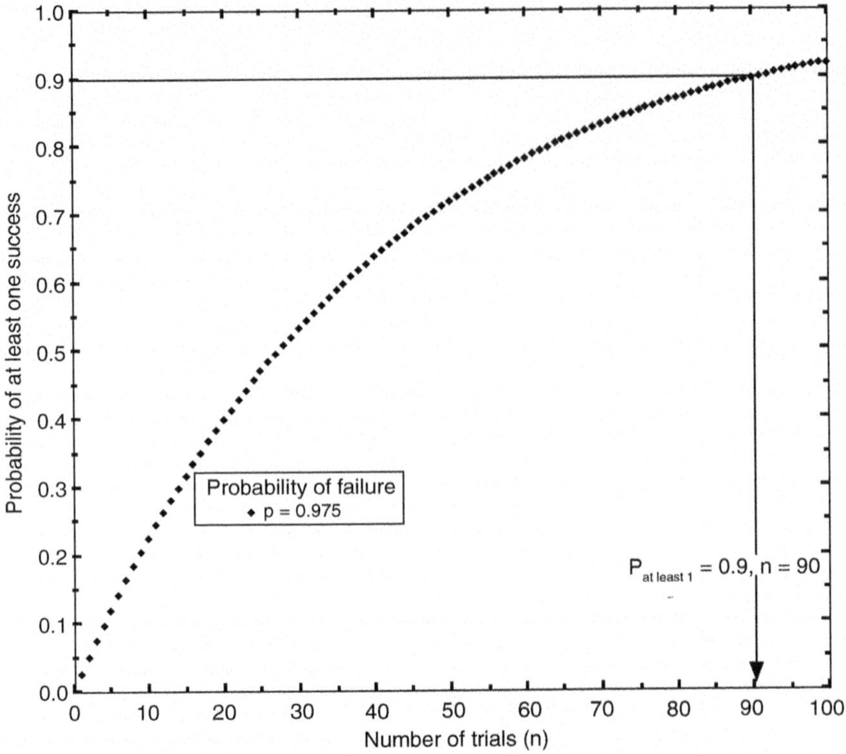

Figure 10.1 Probability of at least one success versus the number of trials. Also shown by an arrow is the number of trials necessary to be 90 percent sure of at least one success, given that the probability of failure for an individual trail is 0.975.

the Middle Tennessee zinc deposit at Elmwood, Tennessee (Callahan, 1977). The number of prospects that must be examined for a fixed probability of at least one success, ($P_{success}$), can be calculated as

$$n = \log(1 - P_{success})/\log(p_{failure}). \qquad (10.2)$$

Risk reduction by increasing the number of prospects examined, whether through submittals from other companies or from internal prospect generation, should give a significant advantage to large firms because only they would have the financial resources to pursue such a policy. Joint venturing, where exploration expenses, responsibilities, and benefits are shared among companies, is the most common way to take advantage of this method of risk reduction, such as discoveries at Yanacocha, Peru, Ladolam, Papua New Guinea, and Lone Tree, Nevada (Sillitoe, 1995). Along with the reduced risk of economic failure comes a reduced financial return in joint ventures. Even with very large resources for exploration, at some point, expenses of exploration can exceed the value of the target found, so there are limits to this method. However, there are other ways to reduce exploration risk that have some advantages over this brute force method.

Risk Reduction by Changing Probability of Failure

It is clear from figure 10.2 that if the probability of failure in a single trial can be reduced, then the probability of at least one success increases dramatically for a given number of trials. Looking for targets that are easier to find can reduce the probability of failure per trial where a trial refers to the examination of a prospect or a deposit that might be economic.

In mineral exploration, discovering easier to find targets means looking for more widely occurring and typically smaller deposits, or looking in places previously unexplored. The problem is that at some point, the deposits are so small that they are not economic to mine. Small deposits and occurrences are relatively numerous, so they would quickly consume available exploration and development money if they were the target (chapter 5). The importance of target size on exploration risk can be shown by the following equation in which the expected amount of metal is estimated from a population of deposits (Singer and Kouda, 1999a):

$$E(\text{metal}) = E(n) \bullet 10^{(\hat{u}\text{tons} + \text{vartons}/2 + \hat{u}\text{grade} + \text{vargrade}/2)} \quad (10.3)$$

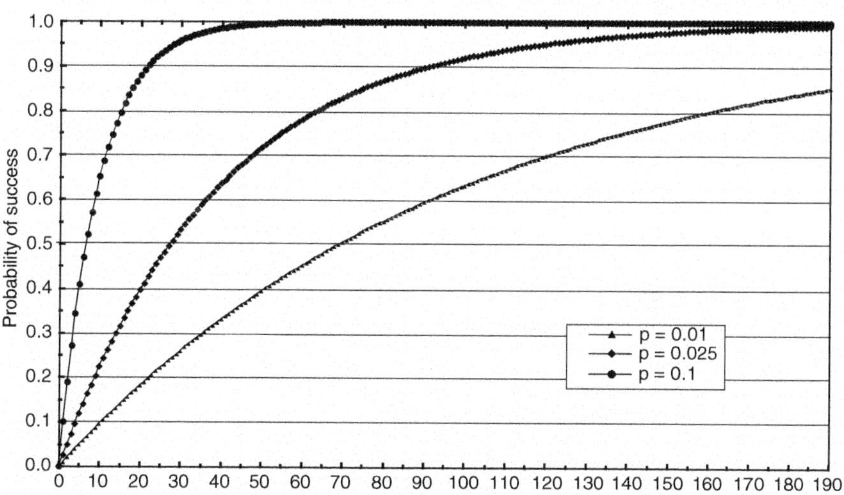

Figure 10.2 Comparison of number of trials versus probability of success at three different probabilities of success per trial.

where E(metal) is the expected amount of metal, E(n) is the expected number of deposits, ûtons is the mean of logged tonnage, vartons is the variance of tonnage, ûgrade is the mean of logged grade, and vargrade is the variance of grade. An important measure of economic success, net present value (chapter 5), depends on contained metal—sensitivity of the expected amount of metal to number of deposits, and their grades and tonnages, can be estimated from equation 10.3. An example using equation 10.3 with porphyry copper deposits is shown in figure 10.3. A 10 percent change in mean grade (base change = 1.1 or 0.9) results in a 55 percent increase (change in expected metal = 1.55) or 35 percent decrease in expected metal content. Expected metal content changes 10 percent when there is a 10 percent change in the expected number of deposits. A 10 percent variation in mean tonnage results in a 650 percent increase (data not shown) or 85 percent decrease in expected metal content (figure 10.3). Variation in mean tonnage is of overwhelming importance in determining metal content. Although

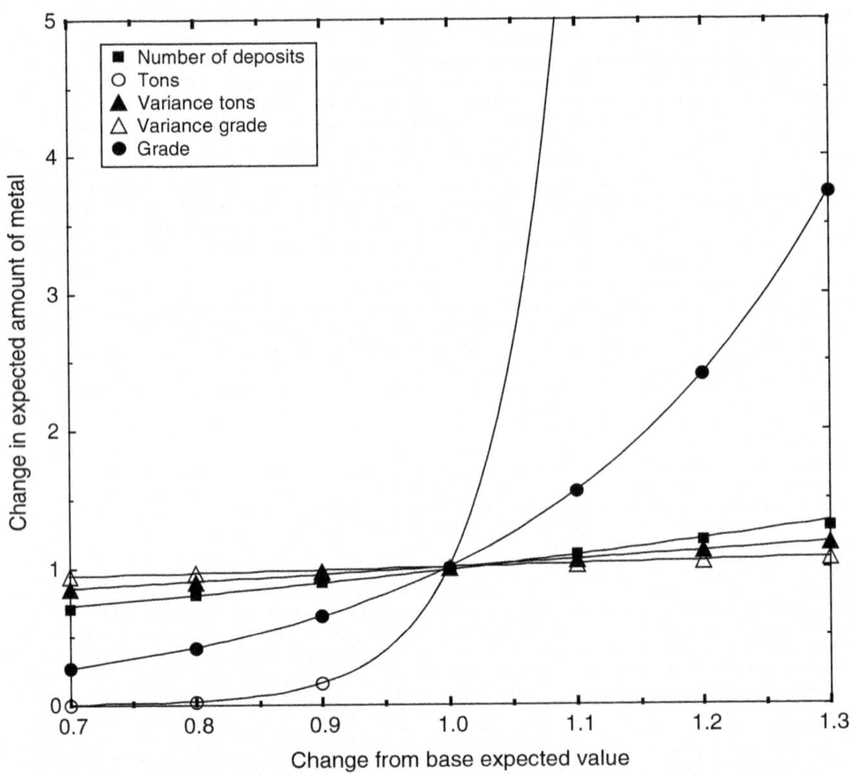

Figure 10.3 Sensitivity of expected amount of copper in porphyry copper deposits with respect to possible changes in expected number of deposits and means and variances of log tonnage and log copper grade.

we have assumed that the distribution of deposit sizes and grades can be represented by independent lognormal distributions in equation 10.3, any distribution that honors the highly skewed nature of deposit tonnages will produce similar results (chapter 2). Based on these results, the greatest opportunity for reducing uncertainty and risk in exploration and resource assessment seems to be by lowering the uncertainty associated with tonnage estimates, followed in importance by uncertainty associated with grade estimates. Exploration enterprises therefore commonly use an economic filter (Penney et al., 2004; see also chapter 5) that is made operational by requiring a minimum size deposit. This points to the principal source of risk in mineral exploration: in order to be economically successful, explorationists seek a rare occurrence in nature, a mineral deposit, and they seek the least common size of mineral deposit, the exceptionally large deposit.

ASSESSMENT EXAMPLE

Assisting industry in making informed decisions on high-risk exploration in poorly explored areas covered by shallow sediments was the main goal of GeoScience in Victoria, Australia, in their assessment of undiscovered gold in the Bendigo Zone (Lisitsin et al., 2007). More than 2,000 metric tons of gold have been produced from an area of about 15,000 km^2 in the central, mostly exposed, part of the Bendigo Zone. The Cenozoic cover effectively masks the prospective 7,600 km^2 Paleozoic rocks of the northern Bendigo Zone and precluded the large-scale early gold prospecting that led to discoveries of the gold deposits in the exposed area to the south. Although there was an estimated 90 percent chance of 15 or more and a 50 percent chance of 24 or more undiscovered deposits, most of the total undiscovered gold is expected to be in deposits that contain more than 31 tons (1 million ounces) of gold. Risk of exploration failure critically depends on the success of finding one of the few largest deposits from the highly skewed distribution of gold in this kind of deposit—critical information in exploration planning (see chapter 6).

Identification of minimum size deposits can be addressed by recognizing and using the significant differences in grades and tonnages among different types of mineral deposits (chapter 6). These differences are clearly demonstrated in figure 10.4, where, if the objective is finding a world-class gold deposit (at least 100 tons of gold), in porphyry Cu, Homestake Au, or Comstock epithermal Au deposits, then porphyry copper deposits require fewer deposits to be examined than either of the deposit types noted for their gold, all other things being equal. Similar differences of the proportion of world-class deposits exist among deposit types for other metals (tables 10.1 through 10.5).

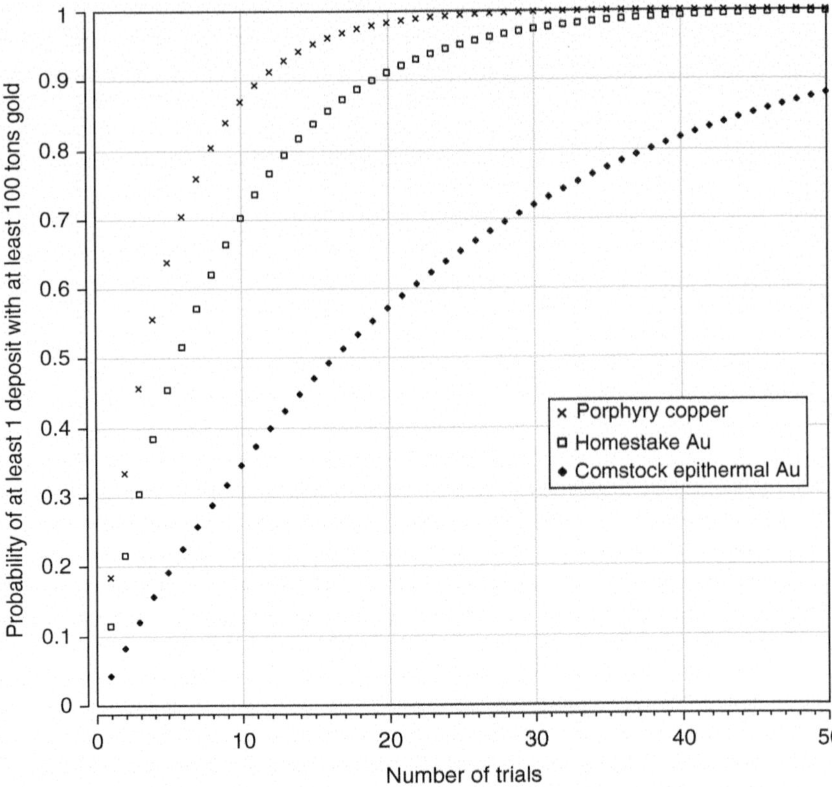

Figure 10.4 Probability of at least one deposit containing at least 100 tons of gold plotted against number of deposits needed to be examined by deposit type.

Because of the high correlation between tonnage of ore and tonnage of contained metal in deposit types (Boldy, 1977; Singer, 1995) and the common association of lead and zinc, many world-class lead deposits are also world-class zinc and perhaps world-class silver deposits. Thus, although the goal may be stated in terms of one contained metal, the chances of success can be affected by searching for different deposit types also affecting where exploration takes place because of the linkage between deposit types and geologic settings. This strategy was key to Homestake's success in finding the McLaughlin hot-spring gold deposit (Anderson, 1982).

These chances must be weighed against the profitability of different types, abilities of the searchers, number of undiscovered deposits that might exist, areas available for search, and regional or local differences within types. One might choose to search for Mississippi Valley Zn-Pb districts or kuroko massive sulfide deposits of which perhaps 90 percent of known deposits or districts have been economic, before searching for

Table 10.1. Percent of world-class gold deposits by deposit type.

Deposit Type	Number of Deposits	Percent Deposits >100 t gold
Epithermal quartz alunite Au	24	29
Sediment-hosted Au	48	17
Porphyry Cu	422	20
Hot-spring Au-Ag	27	11
Comstock epithermal vein	166	4
Homestake Au	243	11
Creede epithermal vein	31	6
Low-sulfide Au-quartz vein	413	6
Replacement-skarn Zn-Pb	147	1
Kuroko massive sulfide	421	2

Table 10.2. Percent of world-class copper deposits by deposit type.

Deposit Type	Number of Deposits	Percent Deposits $>2 \times 10^6$ t Cu
Sediment-hosted Cu	141	15
Porphyry Cu	422	32
Kuroko massive sulfide	421	1

Table 10.3. Percent of world-class zinc deposits by deposit type.

Deposit Type	Number of Deposits	Percent Deposits $>1.7 \times 10^6$ t Zn
Mississippi Valley Zn-Pb	111	20
Shale-hosted Zn-Pb	57	39
Replacement-skarn Zn-Pb	147	14
Sandstone-hosted Pb-Zn	22	10
Kuroko massive sulfide	421	7
Creede epithermal vein	31	3

Table 10.4. Percent of world-class lead deposits by deposit type.

Deposit Type	Number of Deposits	Percent Deposits $>1 \times 10^6$ t Pb
Shale-hosted Zn-Pb	57	39
Mississippi Valley Zn-Pb	111	11
Replacement-skarn Zn-Pb	147	16
Sandstone-hosted Pb-Zn	22	18
Kuroko massive sulfide	421	2

Table 10.5. Percent of world-class silver deposits by deposit type.

Deposit Type	Number of Deposits	Percent Deposits >2,400 t Ag
Replacement-skarn Zn-Pb	147	25
Shale-hosted Zn-Pb	57	19
Creede epithermal vein	31	16
Sediment-hosted Cu	141	4
Epithermal quartz alunite Au	22	9
Mississippi Valley Zn-Pb	111	3
Porphyry Cu	422	9
Comstock epithermal vein	166	5
Kuroko massive sulfide	421	6
Cu skarn	70	1
Polymetallic vein	75	1

porphyry copper deposits of which about 50 percent have not been developed. Information from data used to construct grade and tonnage models (chapter 6) is useful in determining chances of different size deposits occurring. If the number of known deposits is proportional to the number yet to be discovered, then one would search for kuroko type deposits with more than 400 discovered, rather than the four times more scarce Mississippi Valley Zn-Pb districts or the less common sediment-hosted gold deposits.

Risk Reduction within Deposit Types

The strategy of focusing exploration only on world-class deposits has the advantage that the risk of economic loss from mining an uneconomic deposit is significantly reduced at the expense of having few or no deposits to examine. That is, there may be no economic world-class deposits remaining to be discovered in a specific exploration setting. There is also the potential loss of deposits that are economic but are not examined because they appear to be smaller than some predetermined size. By increasing economic risk, it is possible to reduce the risk of not finding a mineral deposit. The balancing of economic filters, geologic theory, and the distribution of deposit sizes remaining in an exploration setting provides opportunities for risk reduction (Penney et al., 2004).

Experience in petroleum exploration demonstrates that larger deposits tend to be discovered early in an exploration play (Drew, 1990). Finding larger deposits or pools early reduces the sizes and values of remaining deposits and affects discovery chances because discovery chances are a function of deposit size. In an analysis of petroleum exploration of the Powder River Basin, Wyoming, Drew (1975) showed that some explorationists were able to reduce their risk of failure by about 43 percent by exploring around the discovery of a large deposit. This risk reduction came at a price of finding only 36 percent as much oil per hole.

The only published study on metal-bearing deposits showing a pattern of finding larger deposits early in the exploration process was on mercury deposits in California (Chung, Singer, and Menzie, 1992). Epithermal gold deposits in Nevada and carbonatite deposits in Brazil show no relationship between size and discovery order—in both cases, however, large discoveries were made late in the exploration process in areas of difficult access. In the case of carbonatite deposits in Brazil, one of the largest deposits, Seis Lagos, was discovered in recent years in the remote headwaters of the Amazon River. The larger Nevada epithermal gold deposits discovered in 1890–1910, such as Round Mountain and Goldfield, are located off the paths to California, which is where most Nevada epithermal deposits discovered in 1840–1870 are located.

It has been shown that for some deposit types, such as porphyry copper, the larger deposits should be discovered earlier than smaller deposits (Singer and Mosier, 1981b). However, this is true only within fixed exploration settings such as an exposed permissive rock that has all parts equally accessible. The relationship between the size of mineralized area and chance of discovery was used successfully in the search for the Mississippi Valley-type deposit at Elmwood, Tennessee (Callahan, 1977). When the exploration setting changes, for example, looking under shallow cover with a particular technique, then the process of finding larger deposits starts over—Boldy (1977) demonstrated the effect exploration method on deposit size discovery order in the search for massive sulfide deposits. Based on an analogy with petroleum exploration, one could reduce the risk of exploration failure by following other's discoveries in a new exploration setting but expect to find smaller, perhaps uneconomic deposits.

Alternatively, one could reduce the risk of finding an uneconomic deposit by focusing on frontier exploration areas and taking advantage of the relationship between the size of deposits and the chance to discover them. Boldy (1977) discusses this trade-off in the search for volcanic-hosted massive sulfide deposits in Canada. An example illustrating these kinds of possibilities for porphyry copper deposits is presented in figure 10.5. Because of the strong positive relationship between area of sulfides (disseminated pyrite) and the deposits' contained copper, a large sulfide system would be a good sign that a large porphyry copper deposit might be present. The presence of a very large sulfide system and a strong relationship as demonstrated in figure 10.5 encouraged continuing exploration of the Pebble Copper property until

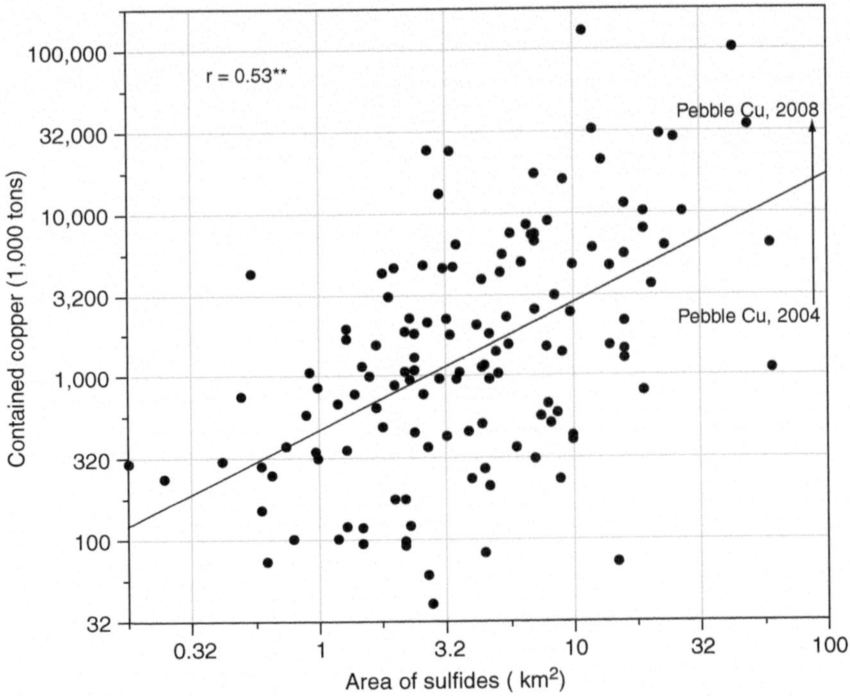

Figure 10.5 Altered area of sulfides (in km²) versus thousands of tons of contained copper in porphyry copper deposits with linear regression line. Change in copper contained in the Pebble Copper deposit between 2004 and 2008 drilling results are indicated by arrow. The correlation coefficient ($r = 0.53$) is significantly different than zero at the 1 percent level.

a very large deposit was discovered by 2008. Another frontier exploration area is under cover.

Large potential rewards motivate exploration for mineral deposits under cover. Increased interest is driven by opportunities to find large deposits. Under cover is the only place remaining to be explored in some regions. Very large deposits are more likely to have high net present values (chapter 5), and they are more likely to be discovered in early stages of exploration (Singer, 2005; Singer and Kouda, 2006). With large rewards come lower chance of discovery, higher exploration costs, and higher risk of economic loss. Capital costs of underground mining can be 70 percent more and operating costs more twice those of surface mining. Net present values can be less than half of the similar deposits on the surface. Higher underground mining costs can be offset by larger deposits, higher commodity prices, or higher grades.

ASSESSMENT EXAMPLE

Although intended to be a prototype for a U.S. national assessment and to aid managers such as the Bureau of Land Management, the greatest benefit of the U.S. Geological Survey's three-part assessment of Nevada may have been reducing exploration risk. Nevada's position as the nation's largest silver producer 130 years ago and as the largest gold producer in 1988 is striking considering that more than 50 percent of Nevada's 286,200 km^2 surface is covered with apparently barren rocks and sediments. Because the majority of mineral deposits exposed at the surface are believed to have already been found, a prime concern of this twelve-chapter assessment prepared by thirteen scientists was the nature of and the depth to possible mineralized systems under this cover. Analysis of regional gravity data was used (Jachens, Moring, and Schruben, 1996) to estimate the thickness of Cenozoic cover and to produce a gravity map from which the effects of thick deposits of young rock and unconsolidated sediments have been removed. This map is also used to help identify the lithology of the concealed basement, to delineate major crustal structures and boundaries, and to identify plutons and concealed calderas, all of which can reflect geologic environments permissive for certain types of mineral deposits. Perhaps the most important result of the gravity analysis related to mineral resources is the conclusion that vast new areas of Nevada may be amenable to exploration for basement-hosted mineral deposits. The previously undocumented extent of basement lying at a depth of less than 1 km represents an important target for mineral exploration and significantly reduced exploration risk (see chapter 7).

Exploration risks under cover can be reduced by using the three parts of assessments and related models and data: (1) grade-and-tonnage and economic filters to provide guides to select appropriate deposit types, (2) descriptive and deposit density models as guides to general location and numbers of deposits, and (3) quantitative properties of mineral deposit sizes, shapes, and orientations of spatially related variables to guide exploration and to reduce location errors.

If a particular size of deposit is being sought, it is possible to determine the permissive tract's area necessary to be assured of a certain probability of a deposit from the equations developed for deposit densities (chapter 4). For example, if a deposit size of 1 million tons is the target of exploration and the decision-maker requires that there be a 90 percent chance of at least one deposit, the permissive tract should be at least 320 km^2 in extent (figure 10.6). If one wants to be assured of at least two deposits containing one million tons, the permissive tract is not twice the area required for one

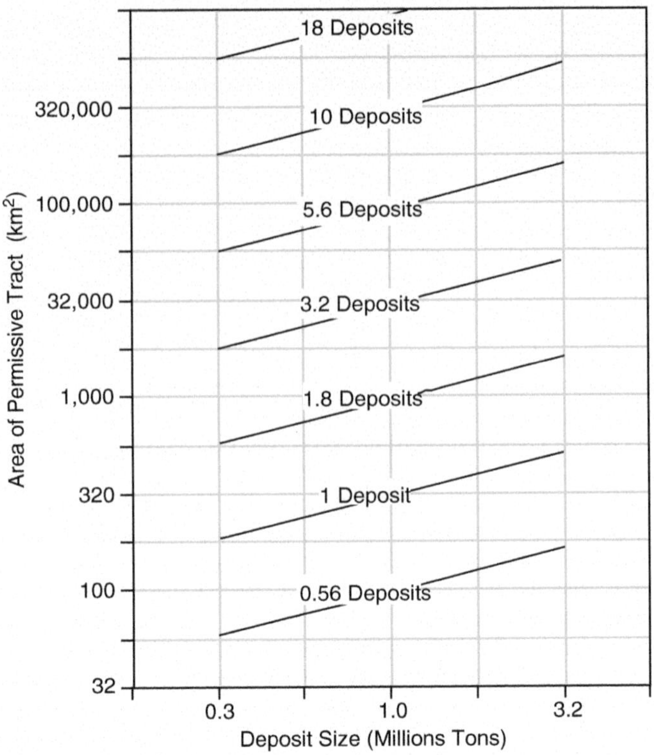

Figure 10.6 Ninety percent confidence contours of number of deposits at various deposit sizes and permissive tract areas. For a 1 million ton deposit, a permissive area of 320 km² is needed to be 90 percent sure of at least one deposit.

deposit, but needs to be about 1,000 km² in extent (figure 10.6). These estimates would only be valid for permissive tracts that are consistent with the deposit models and rules used for permissive tracts as described in chapter 7. Careful applications of the models presented in this book allow improved identification and, ideally, reduction of exploration risks.

Risk Reduction from Political and Security Sources

The risk of loss of investment or personnel due to instability of a government can be the key factor in an exploration decision. As pointed out in an article in *Mining Journal* (1998), in some countries the government has ceased to function or is unable to maintain law and order. Expropriation or security of tenure may be risks in some countries. In some countries, the severance of the right to explore from the right to mine causes risk of loss of exploration investment.

Changes in a government's tax policy can introduce considerable risk to companies if the taxes or ownership rules appear to prevent the company from making a profit. Governments are typically motivated by the desire to get a "fair share" of the peoples' resources, but making the share too large can result in reduced or no future exploration. Historically, when governments have adjusted their expected share to realist levels, companies are slow to return because uncertainties then exist about future policies.

Among the ways to reduce political or security risk are to involve investors who are unlikely to be expropriated, to avoid countries where the risk is high, or to require a high rate of return on an investment. A high rate of return translates into a short payback period—thus shortening the exposure period and, if there are many independent investments, reducing the average risk of loss. Because poverty and the distribution of income are at the root of instability in some countries, requiring a very high rate of return may add to instability by breeding resentment.

Risk Reduction from Economic Sources

Most risk of failure for economic or technical reasons stems from commodity prices being lower than expected, ore reserves being lower than estimated, costs being higher than estimated, and beneficiation difficulties such as poor recovery, currency exchange rates, or delayed development (Hedger, 2008). Acts of nature such as floods also introduce risk, as shown by the drought at the Ok Tedi mine in Papua New Guinea in 1997. Some recommend that risk be adjusted for by increasing the economic return required for an investment. This strategy is suitable only where there are multiple investments with independent risk factors. In mineral exploration and development, other, more specific actions can frequently be taken to reduce risk.

For large low-grade deposits, errors in grade estimates are a major source of risk. Reliability of grade and tonnage ore reserve estimates is typically a function of the amount of information gathered. If actual grades or tonnage are below certain values, the deposit will be uneconomic—that is, there will be economic loss. Drilling more holes both decreases the expected value of the deposit and reduces the uncertainty of the value of the deposit. The marginal benefit of obtaining more information to reduce the risk of a bad investment must be balanced against the costs in money and time of additional information (Mackenzie, 1994). Figure 10.7 can be viewed as a general scheme for representing the trade-offs between gathering more information and reducing uncertainty of an economic return. Unfortunately, some risks, such as those due to price, may not be controllable by gathering more information.

Large variation in commodity prices is common in the mineral industry. This is one reason that world-class gold deposits are sought as exploration targets—world-class deposits tend to be less affected by commodity prices. This helps in the consideration of whether a deposit is economic to mine, but price variation during mining can also close a mine. Variation of prices

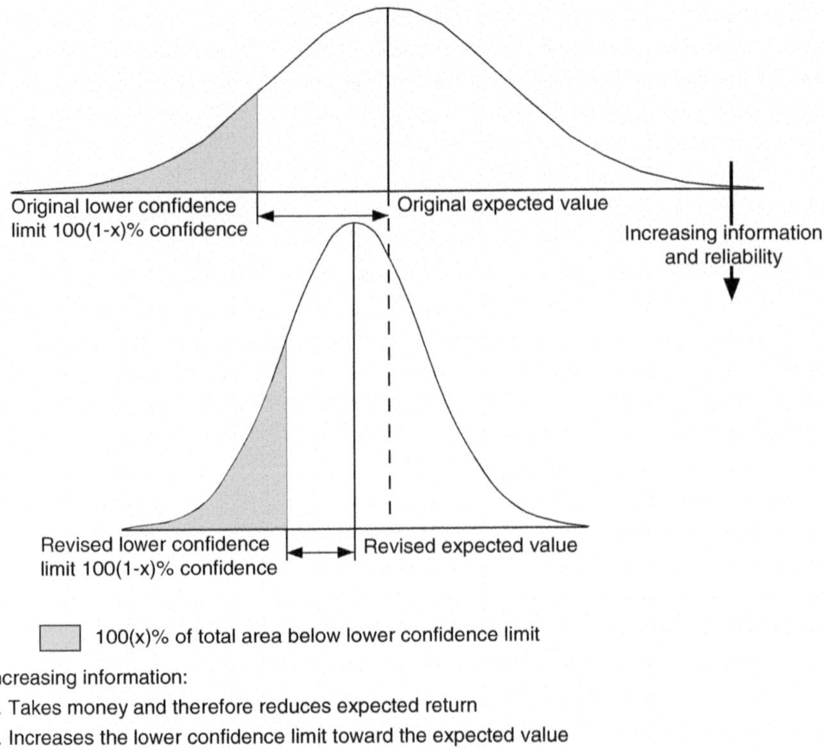

Figure 10.7 Relationship between increasing information and reduced uncertainty and return. After Mackenzie (1994).

during the mine's life can effectively be controlled by hedging. Thus, for some of the entire mine's future output, a contract is agreed to for the sale to a specific customer for a fixed price. In addition to such forward sales, various options contracts may be bought and sold in a hedge strategy. When prices fall after the agreement, the seller is viewed as wise, but the converse is also true—both the buyer and the seller have reduced uncertainty for a price. Another method of reducing risk associated with metal price variability is to seek mineral deposits that contain multiple metals in the hope that when one metal price declines, the other metal prices may move higher. For some periods of time metal prices move independently or even inversely, but, over the long term, many tend to move together, thus mitigating any risk reduction function.

Although not part of exploration in the context used here, it is possible to reduce the risk of loss in mining by sequentially developing and expanding a mine—thus reducing the capital exposed at early stages and reducing the present value of that risked capital. This strategy might also be effective where there is risk of loss due to governmental instability. Some mining and processing methods such as open-pit, heap-leach, or underground shrinkage stope are particularly convenient for this strategy. Because this strategy is easier to employ in some types of mineral deposits than in others, it can play a role in risk reduction in the exploration stage by affecting deposit types sought. Sediment-hosted gold deposits are an example of a type ideal for sequential development—Carlin's mining rate was sequentially increased to 12 times as many tons per year over a 25-year period, and Jerritt Canyon's rate was increased to five times as much over an 11-year period.

Risk Reduction Using Prior Information and Learning

Up to this point, we have assumed that the probability of discovering a deposit is unrelated to the success or failure of previous examinations. Information on the results of early exploration can be used to adjust the views about the existence of a deposit greater than some size. Let D = a deposit of size X or larger exists, d = a deposit of size X or larger does not exist, B = a deposit of size X or larger is found, and b = a deposit of size X or larger is not found. The probability of missing a deposit given that it exists can be considered the risk and be represented as

$$\beta = P(b \mid D). \qquad (10.4)$$

Using Bayes's formula, we can determine the probability that a deposit of size X exists, given that it was not found:

$$P(D \mid B) = P(B \mid D)\, P(D)/(P(B \mid D)\, P(D) + P(B \mid D)\, P(D)) \qquad (10.5)$$

Because $P(b \mid d) = 1$ and $P(d) = 1 - P(D)$, we can simplify equation 10.5 (Gilbert, 1987) to

$$P(D \mid b) = \beta\, P(D) \;/\; (\beta\, P(D) + 1 - P(D)). \qquad (10.6)$$

Figure 10.8 shows the large affects due to changes in the probability of missing a deposit as calculated from equation 10.6. For example, if the prior probability of a deposit existing is 0.5 and the probability of missing is 0.5, equation 10.6 provides a revised probability of existence, given failure to find of 0.33. But if the conditions are the same except that the probability of failure is 0.1, the revised probability of existence becomes 0.09. Revisions of existence probabilities are most noticeable where the prior probability is greater than 0.5 (figure 10.8). Now, with equation 10.6 and figure 10.8, we show the effects of learning about the success or failure of past trials and changing the probability of the existence of a deposit with that information. This, in turn, would change the probability of detection of a deposit in the next trial.

In equation 10.1 where we examined the basics of risk, we made an assumption that the results of one coin toss have no effect on the next coin

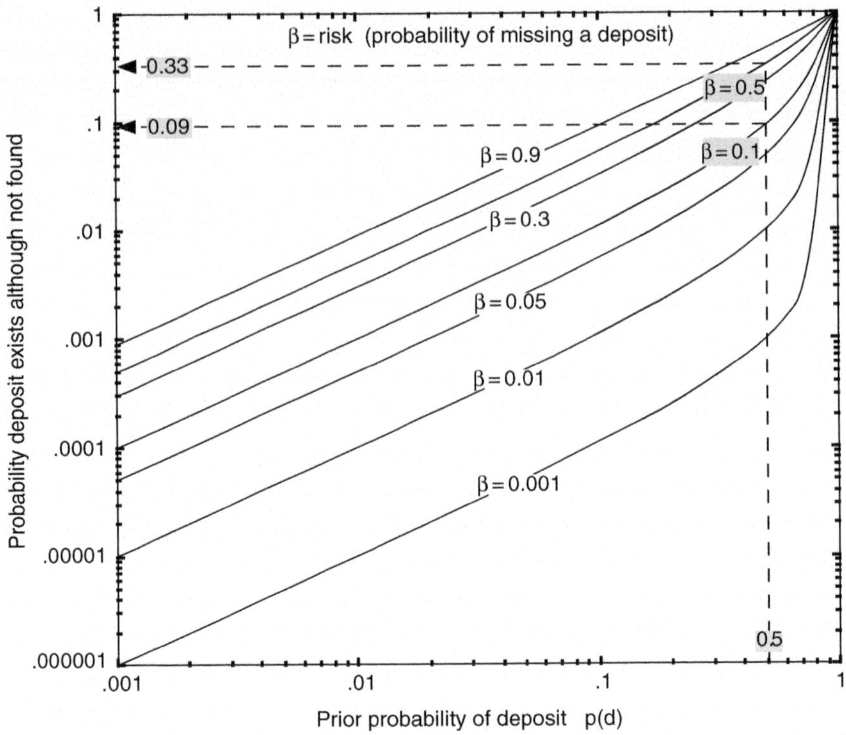

Figure 10.8 Probability that a deposit exists given that it has not been found given different prior probabilities and probabilities of missing a deposit.

toss—this assumption allowed us to examine exploration risk reduction by brute force. There is another form of learning where understanding of the geology and deposit model might be modified as exploration progresses. This form of learning, if successfully applied, can have a profound effect on reducing risk in exploration; poorly applied, it can lead to complete failure. Key to successful application of this form of risk reduction is the nature of the exploration organization. Clearly, organizational culture needs to encourage learning and willingness to reject present models if warranted and yet maintain focus. Rose (1992) discusses organizational traits that adversely affect profits in petroleum exploration firms—many involve risk aversion behavior that discourages the very learning that can perhaps most significantly reduce exploration risk.

Summary

Risk in mineral exploration is examined so the sources of risk can be identified and incorporated in the decision-making process in order to reduce controllable risk. Risk is defined as the chance of failure or loss. Exploration is an economic activity involving risk and uncertainty, so risk also must be defined in an economic context. Both expected monetary values and the probabilities of various outcomes such as economic failure should be considered in the decision-making process. Because economic return when measured by net present value (chapter 5) is closely related to the size of mineral deposits, and because deposit sizes are represented by highly skewed frequency distributions, achieving expected monetary or higher values tends to be a low-probability outcome. This and the typical rareness of mineral deposits are the principal reasons for high risk in mineral exploration. Models, data, and assessment guides presented in the chapters of this book can be used effectively to identify and reduce some exploration risks.

Risk reduction focuses on the strategies using geology, economics, and statistics. A fundamental way to reduce risk is by increasing the number of prospects examined, such as used in the discovery of Elmwood, Tennessee (Callahan, 1977). Joint venturing, where exploration expenses, responsibilities, and benefits are shared among companies, is the most common way to take advantage of this method of risk reduction.

A second fundamental way of risk reduction is to reduce the probability of failure per prospect. Balancing of economic filters, geologic theory, and the distribution of deposit sizes remaining in an exploration setting provides opportunities to lower the probability of failure per prospect. The greatest opportunity for reducing uncertainty and risk in exploration and resource assessment lies with lowering the uncertainty associated with tonnage estimates, followed in importance by uncertainty associated with grade estimates. Exploration enterprises therefore commonly use an economic filter that is made operational by requiring a minimum size deposit. This is why world-class mineral deposits are the primary exploration targets of many mining firms. Identification of minimum size deposits can be addressed by recognizing and using significant differences in grades and tonnages among deposit types. For example, a sediment-hosted gold (Carlin type) deposit is much more likely to be world-class than is a low-sulfide quartz-gold vein deposit.

Experience in petroleum exploration demonstrates that larger deposits tend to be discovered early in an exploration play. Finding larger deposits early reduces the sizes and values of remaining deposits and affects discovery chances. Some petroleum explorationists have reduced their risk of failure by using the strategy of exploring around discoveries of large deposits. This risk reduction comes at a price of finding less oil per hole. Based on an analogy with petroleum exploration, one could reduce the risk in mineral exploration by following others' discoveries in a new exploration setting, but the explorationists should expect to find smaller, perhaps uneconomic deposits. Alternatively, one could reduce the risk of finding uneconomic deposits by focusing on frontier exploration areas such as under cover and taking advantage of the relationship between the size of deposits and the chance to discover them.

Most risk of failure for economic or technical reasons stems from commodity prices being lower than expected, ore reserves being lower than estimated, costs being higher than estimated, beneficiation difficulties such as poor recovery, currency exchange rates, or delayed development. Some recommend that risk be adjusted for by increasing the economic return required for an investment. This strategy is suitable only where there are multiple investments with independent risk factors. Variation of prices during the mine's life can effectively be controlled by hedging. Another method of reducing risk associated with variability of metal prices is to seek mineral deposits that contain multiple metals in the hope that when one metal price declines, the other metal prices may move higher. Although not part of exploration, it is possible to reduce the risk of loss in mining by sequentially developing and expanding a mine, thus reducing the capital exposed at early stages and reducing the present value of that risked capital. Some mining and processing methods such as open-pit, heap-leach, and underground stoping are particularly convenient for this. Because this strategy is easier to employ in some types of mineral deposit than in others, it can play a role in risk reduction in the exploration stage by affecting deposit types sought.

The third fundamental way to reduce risk is to use prior information to modify estimates or to change the assumption of independence in equation 10.1, that is, to learn. With Bayes's formula, the effects of learning about the success or failure of past trials can be used to change the probability of the existence of a deposit. This, in turn, would change the probability of detection of a deposit in the next trial.

Other forms of learning are where geologists learn through training or the experience of others or where understanding of the geology and deposit model might be modified as exploration progresses such as in the discovery of McLaughlin (Anderson, 1982). This form of learning, if successfully applied, can have a profound effect on reducing risk in exploration; poorly applied, it can lead to complete failure. Perhaps the most important way to reduce exploration risk is to employ personnel with the appropriate experience and yet who are still learning.

11

The Future of Quantitative Resource Assessments

Perspective

The difference between the ideas presented by Allais (1957) fifty years ago and those presented in this book reflect a significant growth in knowledge since his work, and the recognition of the value of, and ways to capture, geologic information. We now can use geologic maps to divide large regions into parts that could contain different kinds of mineral deposits, and we know that these different kinds of mineral deposits are significantly different in the amounts and qualities of minerals of interest to society, which affect chances that the deposits will be sought, found, and exploited by society. It is important to remember that our goal is to provide unbiased estimates of undiscovered mineral resources and then to minimize the uncertainty associated with the estimates.

Here we point out where there are opportunities to improve the three-part form of quantitative mineral resource assessment. Many of these opportunities come from identified sources on uncertainties in present assessments of all kinds, such as assessing resources under cover. Some of the improvements can be made in parts of the present assessments that are not completed such as economic filters. Additional opportunities come from the possibilities of harnessing the power of new technologies such as probabilistic neural networks to well-designed applications in these kinds of assessments.

Future quantitative assessments will be expected to estimate quantities, values, and locations of undiscovered mineral resources in a form that conveys both economic viability and uncertainty associated with the resources. Uncertainties about undiscovered resources can be addressed and reduced through improved mineral deposit models, better economic filters and simulators, and application of new technologies to integrate information and by better dealing with geographic uncertainty due to covered terrains (Singer, 2001). Finally, all of these possible ways to improve assessments rely on careful applications of the tools.

Research Opportunities Related to Models

Research opportunities in quantitative resource assessment could be identified in at least three ways: (1) by listing unfinished or flawed parts of assessment tools, (2) by pointing to new technologies that could improve assessments, and (3) by focusing on tasks that could most significantly reduce uncertainties in assessments, and here we consider each. First we examine mineral deposit modeling opportunities, including economic filters. Next we consider improvements needed in combining estimates. An overview of some advanced technologies that might improve future assessments is followed by a section on one of the great sources of uncertainty in assessments likely to be more common in the future—covered terranes.

Mineral-deposit models are important in quantitative resource assessments for two reasons: (1) grades and tonnages of most deposit types are significantly different, and (2) deposit types are present in different geologic settings that can be identified from geologic maps. Mineral-deposit models are key in combining geoscience information for delineation. Many descriptive models suffer from not indicating the map scale of geologic observations that may be important in assessments. Thus, there are parts of the general geologic settings section of descriptive models that list attributes that are observable only in very detailed geologic maps, which typically are not available in assessments. These descriptive models can be significantly improved by having map scales associated with attributes explicitly indicated. Attempts should also be made to capture more regional information that is commonly observable in assessments and is relevant to delineating permissive settings for deposit types. Ideally, these improvements can be documented through counts of the frequencies of occurrence of the attributes associated with known deposits—that is, in digital models. These changes will build the needed foundation for improved delineation of deposit types using digital information.

In chapters 2 and 6 we demonstrated how important uncertainties about grades and particularly tonnages of undiscovered deposits are to an assessment. Selecting the correct deposit model is the most important way of controlling these errors because of the dominance of tonnage, and because deposit models are the best-known predictor of tonnage. However, even identifying the correct deposit type leaves a great deal of uncertainty about possible tonnages of undiscovered deposits. Where some deposits have been discovered and drilled in a region, the question about whether the larger deposits have been discovered early arises. With new research it is possible that a convincing case for size-biased sampling can be made. Research also can lead to the construction of unbiased grade-and-tonnage models that take this possible effect into account.

Of great interest to assessors and to those involved in exploration ventures is the possibility of predicting where the largest undiscovered deposits of a type might occur. To date, research on the topic has met with little success. Some regional differences of sizes of deposit types, such as the significantly higher tonnage than the general porphyry copper deposit type in the Eocene tract of Chile (Cunningham et al., 2008), suggest that such patterns exist and might be predictable. The strong relations among median deposit tonnage, number of deposits, and aerial extent of permissive tract (Singer, 2008) suggest that larger deposits require larger geologically permissive settings. It seems likely that success in predicting where larger deposits occur will depend on demonstrating a positive relationship between the size of host environments, such as permissive basins for sediment-hosted deposits and/or structural settings for many deposit types, and sizes of deposits.

People are interested in the larger deposits because the economies of scale make mining large deposits desirable. Our ability to estimate the economics of undiscovered deposits depends on the quality of our economic cost filters (chapter 5). Much of the fundamental work upon which our cost estimates are based is quite dated and in need of testing and upgrading. Long (2009) suggests that the fundamental relationship between deposit size and mining capacity that has been used for many years needs to be modified. Other relations between capacity and capital and operating costs might also benefit from new research.

Determining whether the undiscovered metals might be economic to recover is an important output of most assessments, and grades and tonnages are necessary inputs for estimating economic viability of mineral deposits. As noted in chapter 9, in order to be able to provide decision-makers with information about undiscovered mineral resources in a form demonstrating possible consequences of their decisions, it is necessary to have a general simulator that incorporates economic filters and captures the uncertainty of resource amounts and values. Software that properly combines the grades and tonnages and properly captures their relationships has been available on computer platforms that no longer exist. Results of attempts to modernize these programs and incorporate economic filters are not yet available.

Advanced statistical learning methods such as probabilistic neural networks and other kinds of kernel methods offer the power either to provide guidelines to experts for their estimates or to make the estimates required in assessments. There is abundant evidence that probabilistic neural networks can classify deposits as well as experienced economic geologists (Singer, 2006; Singer and Kouda, 2003). Probabilistic neural networks have been shown to produce unbiased probabilities if properly trained and tested. There is no reason that they could not estimate probabilities of different geologic settings associated with different deposit types. Nonlinear models such as neural networks should be able to efficiently integrate geophysical and geologic variables in delineating mineral terrains. In order to generalize, it will be necessary to train the system using digital geology and geophysics near known deposits of a specific type. It will then be necessary to test the predictions with similar deposits not used in the training. To demonstrate robustness of results, it will be necessary to train and test the system in several different geologic settings of the selected deposit type from around the world.

Research Opportunities Related to Covered Terrains

Delineating locations that are permissive for undiscovered deposits relies on linking of geologic settings of deposit types as identified in models to geologic environments as interpreted from maps (chapter 7). Because of reliance of these assessments on the link between deposit types and geology, assessing mineral resources under cover has a great deal of uncertainty in delineated boundaries and in associated estimates of number of deposits. Attempts at making objective delineations using digital systems have been less than satisfactory because they have not delineated for specific types of deposits, they have delineated occurrences rather than deposits, or they have produced delineations based on some arbitrary scale rather than probabilities. In addition, none of the existing digital methods studied addresses resources under cover. Even the few methods that have been developed for large regions failed in the important area of assessing mineral resources under cover. Research needs to be performed to develop ways to objectively delineate areas permissive for undiscovered mineral resources remotely through cover based on geophysics and extrapolated geology (Hedger, 2008; Porwal, 2007).

Assessments of areas with resources under cover must rely on extrapolation from surrounding areas, new geologic maps of rocks under cover, or analogy with other well-explored areas that can be considered training tracts. Cover has a profound effect on uncertainty and on methods and procedures of assessments because geology is seldom known and geophysical methods typically have smoothed responses. Mineral exploration enterprises have increasingly been looking for undiscovered deposits under the cover of overlying rocks and sediments. The reasons are that, in some parts of the world, under cover is the only place remaining to be explored, and the belief that the opportunity to find very large deposits may be excellent under cover. Very large deposits are needed to sustain supply and are more likely to have high net present values than are small deposits. Even in parts of the world where exploration of exposed rocks is only partially complete, the belief that the larger deposits tend to be found early in the exploration process suggests that the new frontier of exploration under cover increase the chances of discovering world-class deposits—important discoveries under cover such as Olympic Dam, Spence, Esperanza, Toki, Gaby, Mansa Mina, Far Southeast, Cannington, and Pogo reinforce these beliefs.

Deposit models used with local geology can be applied to define permissive tracts by deposit type. Deposit types are present in broad geologic settings that can be identified from geologic maps and the permissive settings can, at least in a broad sense, be projected under cover. Assessments or planning exploration under cover must rely on extrapolation from surrounding areas, new geologic maps of the rocks under cover, and analogy with other well-explored areas to define general belts to explore. Mineral deposit models that capture the quantitative properties of mineral deposits provide information about the sizes, shapes, and orientations of spatially related variables. These variables include spatially related deposits types, alteration zones, geochemical haloes, geophysical responses, and structural settings. These digital deposit models would be constructed from well-explored deposits elsewhere. Relationships of these variables and mapped geology and geophysics at the appropriate scales could be integrated using the advance learning methods mentioned above to delineate under cover.

Regional or metallogenic settings of mineral deposits can provide guides on general locations of deposits under cover if the geology can be projected under cover. Fortunately, some scientists have experience in making geologic maps under cover. The Australians have had to deal with the problem of cover for some years (Wilford, 2000). As might be guessed, however, the methods require detailed geophysics that, because of its expense, can be applied only in relatively small areas. One area of active research that should prove useful in locating possible places of mineralization under cover is structural geology. Many faults can be identified under cover with geophysics—some kinds of faults are related to certain types of mineralization such as porphyry copper deposits (Berger, Drew, and Singer, 1999).

Beyond Digital

Even with the advances already made through quantitative models and advanced statistical methods, and those on the horizon, in our view it is important to keep expert economic geologists involved in the process. Unfortunately, there are cases where equations or advanced statistical, mathematical, or economic methods have been applied where they are not appropriate. Understanding unstated and underlying assumptions and implications of using a method are still critical in assessments. An example where an expert economic geologist would be helpful is in delineating the deeply emplaced central part of the Sierra Nevada Batholith as permissive for porphyry copper deposits that form less than 5 km below the surface. An example where knowledge of applied statistics would help is the use of various forms of the Bayes equation to combine multiple observations or variables in order to make probabilistic estimates—such estimates are commonly biased upward (Singer and Kouda, 1999b). Perhaps the most damaging example is the widespread use of spatial buffers around geologic units in delineating permissive tracts because it is easy. This practice results in large areas that can be demonstrated to be not permissive and, even worse, large permissive areas that are excluded—particularly permissive settings under cover—all for the lack of an expert geologist to interpret the geology.

The decades of experience of expert and objective forecasting in meteorology provide insight into how the process of making assessments of undiscovered mineral resources might be improved. Murphy and Winkler (1984) found that consensus schemes performed better than almost all individual forecasters and that the best forecasts were made when objective (i.e., computer generated) forecasts were part of the information supplied to expert forecasters. We should always use the best possible methods of making quantitative assessments—it would be unwise to leave the human expert out of the final product.

Appendix 1: Conversion of Units

1 pound (abbr. lb) = 0.45359 kilograms
1 troy ounce (abbr. oz) = 31.103 grams
1 grain = 0.0648 grams
1 metric ton, or tonne (abbr. t) = 1,000 kilograms = 2,204.6 pounds = 1.1023 short tons
1 short ton (abbr. st) = 2,000 pounds = 0.907185 metric ton
1 long ton (abbr. lt) = 2,240 pounds = 1,016.0 kilograms
1 mile (statute) = 5,280 feet = 1,760 yards = 1.6093 kilometers
1 nautical mile = 1.852 kilometers
1 kilometer (abbr. km) = 1,000 meters = 0.62139 miles
1 square kilometer (abbr. km^2) = 0.386100 square miles
1 yard = 3 feet = 0.9144 meters
1 foot (abbr. ft) = 0.3048 meter
1 square foot = 0.0929 square meters
1 acre = 4,047 square meters
1 hectare = 2.5 acres
1 cubic yard (abbr. yd^3) = 0.7646 cubic meters
1 flask (of mercury) = 34.473 kilograms or 76 pounds
1 carat (gem stones) = 0.2 grams
1 U.S. gallon = 3.785 liters
1 imperial gallon = 4.546 liters
1 barrel = 42 U.S. gallons
1 inch on a 1:250,000-scale map = 4 miles (approx.)

To convert short tons to metric tons, multiply short tons by 0.907185.

To convert ounces/short ton to grams/metric ton, multiply oz/st by 34.285 (i.e., 31.103 g/0.907185 st).

To convert ounces/long ton (Australian usage) to grams/metric ton, multiply oz/lt by 30.61 (i.e., 31.103 g/1.0164 lt).

Appendix 2: Algorithm for Estimating Probabilities for a Set of Number-of-Deposit Estimates

Here we reprint the code for the algorithm used for estimating probabilities of number of deposits and expected number of deposits, an extension of the Root, Menzie, and Scott (1992) method for three estimates of number of deposits to five estimates.

```
P[n] is the probability of n deposits
N9, N5, N1, N05, and N01 are user-supplied number of deposit
estimates at the 90th, 50th, 10th, 5th, and 1st percentiles
S1, S2, S3, S4, and S5 are values used in calculations
n is the number of deposits
expect is the expected (mean) number of deposits
  S1 = 1 + 2 * N9
  if N9 = N5 then S2 = 1
  else S2 = 2 + 2 * (N5-N9 - 1)
  if N = N1 then S3 = 1
  else S3 = 2 + 2 * (N1-N5 - 1)
  if N1 = N05 then S4 = 1
  else S4 = 2 + 2 * (N05-N1 - 1)
  if N01 = N05 then S5 = 1
  else S5 = 2 + 2 * (N01-N05 - 1)
  sum = 0.0
  expect = 0.0
  for n = 0 to N01 do
    begin
      if (n < N9) then P[n] = 0.2/S1
```

```
else
  begin
    if (n = N9) then
      begin
        P[n] = 0.1/S1 + 0.4/S2
        if (N9 = N5) then
          begin
            P[n] = P[n] + 0.4/S3
            if N5 = N1 then P[n] = P[n] + 0.05/S4
            if N5 = N05 then P[n] = P[n] + 0.04/S5
          end
        end
    else
      begin
        if (n < N5) then
          begin
            P[n] = 0.8/S2
            if N9 = N5 then P[n] = 0.8/S3
          end
        else
          begin
            if n = N5 then
              begin
                P[n] = 0.4/S2 + 0.4/S3
                if N5 = N1 then
                  begin
                    P[n] = P[n] + 0.05/S4
                    if N1 = N05 then P[n] = P[n] + 0.04/S5
                  end
              end
            else
              begin
                if n < N1 then
                  begin
                    P[n] = 0.8/S3
                    if N5 = N1 then P[n] = 0.0
                  end
                else
                  begin
                    if (n = N1) then
                      begin
                        P[n] = 0.4/S3 + 0.05/S4
                        if N1 = N05 then P[n]=P[n] + 0.04/S5
                      end
                    else
                      begin
                        if n < N05 then P[n] = 0.1/S4
                        else
                          begin
```

```
                                    if n = N05 then P[n] =
                                         0.05/S4 + 0.04/S5
                                  else
                                    begin
                                      if n < N01 then P[n] = 0.08/S5
                                        else
                                        begin
                                        if n = N01 then
                                          begin
                                            P[n]=0.08/S5+ 0.01
                                            if N01 = N05 then
                                            P[n] = P[n] + 0.025
                                          end
                                        end
                                    end
                                end
                            end
                        end
                    end
                end
            end
        end
    end
  for n = 0 to N01 - 1 do
    begin
      sum = sum + P[n]
    end
  P[N01] = 1.0 - sum
  for n = 1 to N01 do
    begin
      expect = expect + P[n] * n
    end
```

Glossary

Accuracy is the closeness of a measurement or estimate to the true value.

Analysis of variance is a statistical method of testing the equivalence of mean values from more than one group or class.

Anchoring is the tendency of an individual to use and adjust from an initial value in making an estimate.

Anomalous describes a group of samples that significantly differ from others in the group.

Background is a group of values characteristic of the most common samples in a population—typically nonmineralized.

Bayesian estimate is the use of Bayes's theorem in which a prior probability can be modified or updated by new information to produce a revised or posterior estimate.

Bias is a deviation of the value or estimate from the true value, for example, when an estimator such as a mean significantly over- or underestimates the true value. A biased sample or estimate is not accurate. Bias occurs when an estimate does not follow normative statistical or logical rules.

Bimodal distribution is a frequency distribution that has two humps or modes. It is typically two distributions combined.

Block caving is an underground mining method used for (1) large ore bodies, (2) massive ore bodies that have large vertical extension, and (3) rock that will cave and break into manageable masses.

Capital expenditure is money invested in the development of an operation.

Cash flow is net expenditures and income in a period of time (usually a year).

Censored data are measurements for some population where some units are not available, such as those not reported, or reported as trace, or not detected.

Coefficient of variation is the standard deviation divided by the mean. It is sometimes expressed as a percent and can be considered a relative standard deviation.

Cover is typically younger sediments or rocks that are over older mineral deposits. The term can also be applied to older rocks or sediments that are tectonically placed over rocks that host mineral deposits.

Crustal abundance is the concentration of an element in the earth's crust. It can also represent the concentration of a chemical compound, such as salt, or a mineral, such as emerald.

Cutoff grade is the lowest grade of a material that can be included into a resource estimate. Mineral resource estimates may include material below the selected cutoff grade to ensure that the mineral resources comprise bodies of mineralization of adequate size and continuity to show reasonable prospects for eventual application of a feasible mining method.

Decision analysis is the application of structured procedures for choosing optimal decisions in the face of uncertainty. The procedure involves breaking the complex problem into tractable parts.

Degrees of freedom refers to the number of samples minus the number of parameters estimated. For the strength of a linear relationship between two variables, the degrees of freedom are the number of samples minus 2.

Dematerialization is an economic term about the relative or absolute reduction of the quantity of materials required to serve some function in society.

Deposit density model is the frequency distribution of number of deposits from the grade and tonnage model per unit of permissive control area. It is commonly applied in a regression of number of deposits versus area permissive. It is used to estimate number of deposits or as a guide for making estimates.

Disaggregation is breaking the problem into component parts to make it easier to solve.

Discounted cash flow is the cash flow for a number of years adjusted to one point in time by a discount factor such as an interest rate.

Discriminatory could mean either necessary or sufficient.

Essential refers to a situation in which, if the evidence is false (event did not happen), then the probability of a deposit is zero.

Expected value is the first moment of a probability density function that corresponds to the mean. It is the most common measure of central location of a random variable.

Expert is a person who has a background and experience in the subject matter at the desired level of knowledge and is viewed as qualified to make estimates or provide advice on the topic of interest.

Expert judgment refers to judgments, advice, or estimates by subject-matter experts.

Exploration play is the exploration for a cluster of petroleum accumulations.

Gambler's ruin is the fact that a gambler will almost certainly go broke in the long run against an opponent with much more money, even if the opponent's advantage on each turn is small.

GDP is gross domestic product.

GIS is geographic information system

Grid is the systemic spatial array of points, samples, lines, or drill holes. Commonly the grid is represented by a square, rectangular, or hexagonal array.

Independent events are events that have no influence on each other.

Inferred mineral resource is that part of a mineral resource for which the overall tonnages, grades, and mineral contents can be estimated with a reasonable level of confidence. It is based on geologic evidence and apparent geologic and grade continuity after applying economic parameters. It is derived from information gathered through appropriate techniques from locations such as outcrops, trenches, pits, workings, and drill holes, and is limited or of uncertain quality and reliability in some way. An inferred mineral resource has a lower level of confidence than that applying to an indicated mineral resource.

Interval scale is a scale of measurement of a variable that uses numbers with a fixed unit of measure to identify the attribute.

Lognormal distribution is a skewed distribution that can be transformed into a normal distribution by taking the logarithm of the values.

Mean is the numeric average of a set of values calculated by summing the values and dividing by the number of values. For distributions skewed to the right, the mean plots to the right of the mode (highest frequency) and to the right of the median (50th percentile).

Median is the value above and below which half the population lies (50th percentile).

Metal endowment is the sum of metal in all occurrences with specified characteristics, such as concentration, size, and depth (Harris, 1984).

Mineral deposit is a mineral occurrence of sufficient size and grade that it might, under the most favorable of circumstances, be considered to have economic potential.

Mineral deposit type is a group of deposits sharing a relatively wide variety and large number of attributes (Cox, Barton, and Singer, 1986).

Mineral occurrence is a concentration of a mineral (usually, but not necessarily, considered in terms of some commodity, e.g., copper, barite, or gold) that is considered valuable by someone somewhere or that is of scientific or technical interest. Typically, exploration is incomplete and information is sparse regarding mineral occurrence (Cox, Barton, and Singer, 1986).

Mineral prospect is a mineral occurrence that has been drilled or investigated in some detail and is believed to have a moderate or small chance of becoming economically viable.

Mineral resource is a concentration or occurrence of material of economic interest in or on the earth's crust in such form, quantity, and quality that there are reasonable prospects for eventual economic extraction.

Model is a simplified representation of a natural reality.

Monte Carlo simulation is a computer method of randomly sampling probability distributions so that they can be combined and the variables' properties can be investigated.

Necessary refers to a situation in which, if the evidence is false (something does not exist), then the probability of a deposit decreases.

Net present value is a measure of the value of an investment after discounting the cash flow at a rate of return and subtracting an initial capital investment.

Nominal scale is a scale of measurement of a variable that uses labels or names to identify the attribute.

Normal distribution is a probability distribution function that is symmetrical about the mean, which is the same as the median and the mode. The mean is independent of the variance.

Ordinal scale is a scale of measurement of a variable that uses the order or rank of nominal data as a measure to identify the attribute.

Parameter is a fixed characteristic of a probability distribution such as the population mean. Parameters are known only when all samples of the population are available. In most practical cases, parameters are estimated and should be called estimates.

Percentile is the value that divides the area of a frequency distribution into specific percentages. For example, the 10th percentile is the value of a distribution such that 10 percent of the distribution is smaller and 90% of the distribution is larger.

Permissive tract is a geographic area delineated such that the probability of deposits of the type delineated occurring outside the boundary is negligible (i.e., <1 in 100,000–1,000,000).

Poisson distribution is a discrete probability distribution that is often useful in estimating the number of uncommon events such as mineral deposits.

Poisson process is a process in which events occur randomly and independent of each other at a fixed rate in time or space.

Population is the entire theoretical or existing set of items under study.

Posterior probability is a probability that has been revised by additional information.

Precision is a measure of the closeness of agreement among individual measurements, for example, the variability about the mean of a group of sample values.

Prediction interval is an interval estimate for an individual value.

Prior probability is an initial estimate of the probability of an event.

Probability is a number between 0 and 1 inclusive that represents the chance of an event. It is commonly thought of as the relative frequency of the occurrence of an event.

Probability distribution is a description of the distribution of probabilities for different values of a random variable.

Probable (indicated) reserves are reserves for which quantity and grade and/or quality are computed from information similar to that used for proven (measure) reserves, but the sites for inspection, sampling, and measurement are farther apart or are otherwise less adequately spaced. The degree of assurance, although lower than that for proven (measured) reserves, is high enough to assume continuity between points of observation.

Proven (measured) reserves are reserves for which (1) quantity is computed from dimensions revealed in outcrops, trenches, workings, or drill holes and grade and/or quality are computed from the results of detailed sampling, and (2) the sites for inspection, sampling, and measurement are spaced so closely and the geologic character is so well defined that size, shape, depth, and mineral content of reserves are well established.

Qualitative assessment is an assessment where the product is presented as labels or names.

Quantile is a number that partitions the values of a frequency into n classes containing the same number of data. The quantile that divides a frequency distribution into two classes is the median.

Quantitative assessment is an assessment where the product is presented as numerical values.

Random variable is a numerical result that has an associated probability distribution and is from an experiment.

Range is the difference between the maximum and minimum values of a sample and depends on the number of samples taken.

Ratio scale is a scale of measurement of a variable that uses numbers as a unit of measure to identify the attribute and in which the ratio of two values is meaningful.

Reserve is that part of a mineral deposit that could be economically and legally extracted or produced at the time of the reserve determination.

Resource is a concentration of naturally occurring solid, liquid, or gaseous material in or on the earth's crust in such form that economic extraction of a commodity from the concentration is currently or potentially feasible.

Risk is the probability and the associated cost of an undesirable outcome (ISO standard).

Robust refers to a model, estimate, or method that tends to produce correct prediction or estimates regardless of error in or problems with the data.

Room-and-pillar mining is removing ore from rooms while leaving pillars in place to support the roof of the room.

Sampled population is the set of population units available for measurement.

Sensitivity analysis is a set of techniques used for assessing the relative importance of model input factors.

Shrinkage stope mining is mining upward, creating a sloping underground room.

Skewed describes a frequency distribution that is not symmetric. Skewed right means the tail is on the right. Skewed distributions are common for variables measuring length, volume, mass, or value.

Subjective probability is an estimate of the chances of an event made by an individual.

Sufficient refers to a situation in which, if the evidence is true (does exist), then the probability of a deposit increases.

Target population is the population for which inferences will be made.

t test is a statistical test of the equality of group means. It is based on the *t* distribution and is commonly used whenever the population standard deviation is unknown.

Type I error is the rejection of the null hypothesis when it is true. The risk of this is the level of significance.

Type II error is the acceptance of the null hypothesis when it is false. The probability of making a Type II error depends on the alternative value and its distribution.

Uncertainty is lack of knowledge of the outcome or result. In many cases it can mean the variability of outcome.

Variable is a quantity that can take on any one of a given set of values.

References

Agterberg, F.P., 1977, Frequency distributions and spatial variability of geological variables, *in* Ramani, R.V., ed., Application of computer methods in the mineral industry: Society of Mining Engineers of American Institute of Mining, Metallurgical, and Petroleum Engineers, New York, pp. 287–298.

Agterberg, F.P., 1984, Discrete probability distributions for mineral deposits in cells: Proceedings, 27th International Geological Congress, v. 20, VNU Science Press, Utrecht, The Netherlands, pp. 205–225.

Aitchison, J., and Brown, J.A.C., 1963, The lognormal distribution: Cambridge University Press, London.

Allais, M., 1957, Method of appraising economic prospects of mining exploration over large territories—Algerian Sahara case study: Management Science, v. 3, no. 4, pp. 285–347.

Anderson, J.A., 1982, Gold—its history and role in the U.S. economy and the U.S. exploration of Homestake Mining Co.: Mining Congress Journal, January, pp. 51–58.

Australian Geological Survey Organisation, ed., 1998, Exploration models for major Australian mineral deposit types: AGSO Journal of Australian Geology and Geophysics (special volume), v. 17, no. 4, pp. 69–258.

Aven, T., 2003, Foundations of risk analysis—a knowledge and decision-oriented perspective: John Wiley & Sons, Chichester, U.K.

Bárdossy, G., and Fodor, J., 2004, Evaluation of uncertainties and risks in geology—new mathematical approaches for their handling: Springer, Heidelberg.

Barnett, H., and Morse, C., 1963, Scarcity and growth: the economics of natural resource availability: Johns Hopkins University Press for Resources for the Future, Baltimore, Md.

Barton, P.B., Jr., 1983, Unconventional mineral deposits—a challenge to geochemistry, *in* Shanks, W.C., III, ed., Cameron volume on unconventional mineral deposits: NewYork, Society of Mining Engineers of the American Institute of Mining, Metallurgical, and Petroleum Engineers, pp. 3–14.

Barton, P.B., Jr., 1993, Problems and opportunities for mineral deposit models, *in* Kirkham, R.V., Sinclair, W.D., Thorpe, R.I., and Duke, J.M., eds., Mineral Deposit Modeling: Geological Association of Canada Special Paper 40, pp. 7–13.

Bawiec, W.J., and Spanski, G.T., 2009, EMINERS Version 2.0.0.10—quick-start guide: U.S. Geological Survey Open File Report 04–1344. Available at: http://pubs.usgs.gov/of/2004/1344.

Beikman, H.M., Hinkel, M.E., Frieders, T., Marcus, S.M., and Edward, J.R., 1983, Mineral surveys by the Geological Survey and the Bureau of Mines of Bureau of Land Management wilderness study areas, U.S. Geological Survey Circular 901.

Berger, B.R., 1986, Descriptive model of low-sulphide Au-quartz veins, *in* Cox, D.P., and Singer, D.A., eds, Mineral deposit models, U.S. Geological Survey Bulletin 1693, p. 239.

Berger, B.R., Drew, L.J., and Singer, D.A., 1999, Quantifying mineral-deposit models for resource assessment, *in* Ódo, L., Korpás, L., McCammon, R.B., and Hofstra, A.H., eds., Deposit modeling and mining-induced environmental risks: Geologica Hungarica Series Geologica, v. 24, pp. 41–54.

Blakely, R.J., Schruben, P.G., and Moring, B.C., 1996 Shallow magnetic lithologies as interpreted from low-altitude aeromagnetic data: Nevada Bureau of Mines and Geology Open File Report 96–2, chap. 3.

Bliss, J.D., 1986, Grade and tonnage model of low-sulphide Au-quartz veins, *in* Cox, D.P., and Singer, D.A., eds., Mineral Deposit Models: U.S. Geological Survey Bulletin 1693, pp. 239–243.

Bliss, J.D., ed., 1992a, Developments in mineral deposit modeling: U.S. Geological Survey Bulletin 2004.

Bliss, J.D., 1992b, Grade–tonnage and other models for diamond kimberlite pipes: Nonrenewable Resources, v. 1, no. 3, pp. 214–230.

Bliss, J.D., and Cox, D.P., 1986, Grade and tonnage models of polymetallic veins, *in* Cox, D.P., and Singer, D.A., eds., 1986, Mineral deposit models: U.S. Geological Survey Bulletin 1693, pp. 125–129.

Bliss, J.D., and Jones, G.M., 1988, Mineralogic and grade–tonnage information on low-sulfide Au-quartz veins: U.S. Geological Survey Open File Report 88–299.

Bliss, J.D., and Menzie, W.D., 1993, Spatial mineral-deposit models and the prediction of undiscovered mineral deposits, *in* Kirkham, R.V., Sinclair, W.D., Thorpe, R.I., and Duke, J.M., eds., Mineral deposit modeling: Geological Association Canada Special Paper 40, pp. 693–706.

Bliss, J.D., Menzie, W.D., Orris, G.J., and Page, N.J, 1987, Mineral deposit density—a useful tool for mineral-resource assessment, *in* Sachs, J.S., ed., U.S.

Geological Survey Research of Mineral Resources—1987 Program and Abstracts: U.S. Geological Survey Circular 995, p. 6.

Bliss, J.D., Orris, G.J., and Menzie, W.D., 1987, Changes in grade, volume and contained gold during the mining life-cycle of gold placer deposits: CIM Bulletin, v. 80, no. 903, pp. 75–79.

Boldy, J., 1977, (Un)certain exploration facts from figures: Canadian Mining and Metallurgical Bulletin, v. 70, no. 781, pp. 86–95.

Brinck, J.W., 1967, Note on the distribution and predictability of mineral resources: Euratom 3461, Brussels.

Brooks, D.B., 1976, Mineral supply as a stock, *in* Vogely, W.A., ed., Economics of the mineral industries, 3rd ed.: American Institute of Mining, Metallurgical, and Petroleum Engineers, New York, pp. 127–207.

Callahan, W.H., 1977, The history of the discovery of the zinc deposit at Elmwood, Tennessee, concept and consequence: Economic Geology, v. 72, no. 7, pp. 1382–1392.

Camm, T.W., 1991, Simplified cost models for prefeasibility mineral evaluations: U.S. Bureau of Mines Information Circular 9298.

Camm, T.W., 1994, Simplified cost models for prefeasibility mineral evaluations: Mining Engineering, v. 46, no. 6, pp. 559–562.

Carranza, E.J.M., and Hale, M., 2002, Where are porphyry copper deposits spatially localized? A case study in Benguet Province, Philippines: Natural Resources Research, v. 11, no. 1, pp. 45–59.

Chung, C.F., Singer, D.A., and Menzie, W.D., 1992, Predicting sizes of undiscovered mineral deposits—an example using mercury deposits in California: Economic Geology, v. 87, no. 4, pp. 1174–1179.

Churchman, C. West, 1968, The systems approach: Dell Publishing, New York.

COMRATE, 1975, Resources of copper, *in* Mineral resources and the environment: National Academy of Sciences, Washington, D.C., pp. 127–183.

Cox, D.P., ed., 1983a, U.S. Geological Survey—INGEOMINAS mineral resource assessment of Colombia: ore deposit models: U.S. Geological Survey Open File Report 83–423.

Cox, D.P., ed., 1983b, U.S. Geological Survey—INGEOMINAS mineral resource assessment of Colombia: additional ore deposit models: U.S. Geological Survey Open File Report 83–901.

Cox, D.P., 1985, Geology of the Tanama and Helecho porphyry copper deposits and their vicinity, Puerto Rico: U.S. Geological Survey Professional Paper 1327, p. 59.

Cox, D.P., 1986a, Descriptive model of porphyry Cu-Au, *in* Cox, D.P., and Singer, D.A., eds., Mineral deposit models: U.S. Geological Survey Bulletin 1693. p. 110.

Cox, D.P., 1986b, Descriptive model of Porphyry Cu, *in* Cox, D.P., and Singer, D.A., eds., Mineral deposit models: U.S. Geological Survey Bulletin 1693, p. 76.

Cox, D.P., 1993, Estimation of undiscovered deposits in quantitative mineral resource assessments—examples from Venezuela and Puerto Rico: Nonrenewable Resources, v. 2, no. 2, pp. 82–91.

Cox, D.P., Barton, P.R., and Singer, D.A., 1986, Introduction, *in* Cox, D.P., and Singer, D.A., eds., Mineral deposit models: U.S. Geological Survey Bulletin 1693, pp. 1–10.

Cox, D.P., Berger, B.R., Ludington, S., Moring, B.C., Sherlock, M.G., Singer, D.A., and Tingley, J.V., 1996, Delineation of mineral resource assessment tracts and estimation of number of undiscovered deposits in Nevada, *in* Singer, D.A., ed., An analysis of Nevada's metal-bearing mineral resources: Nevada Bureau of Mines and Geology Open File Report 96–2, pp. 12.1–12.25. Available at: ftp://ftp.nbmg.unr.edu/NBMG/nbmg_ofr96_2/.

Cox, D.P., Lindsey, D.A., Singer, D.A., and Diggles, M.F., 2003, Sediment-hosted copper deposits of the world—deposit models and database: U.S. Geological Survey Open File Report 03–107. Available at: http://geopubs.wr.usgs.gov/open-file/of03–107/.

Cox, D.P., and Singer, D.A., eds., 1986, Mineral deposit models: U.S. Geological Survey Bulletin 1693.

Cox, D.P., and Singer, D.A., 1992, Distribution of gold in porphyry copper deposits, *in* DeYoung, J.H., and Hammerstrom, J.M., eds., Contributions to commodity research: U.S. Geological Survey Bulletin 1877, pp. C1–C14. (Also available as Open File Report 88–46.)

Craddock, C., 1972, Geologic map of Antarctica: American Geographical Society, New York.

Cunningham, C.G., Singer, D.A., Zappettini, E.O., Vivallo, S.W., Celada, C.M., Quispe, J., Briskey, J.A., Sutphin, D.M., Gajardo, M.M., Diaz, A., Portigiliati, C., Berger, V.I., Carrasco, R., and Schultz, K.J., 2007, A preliminary quantitative resource assessment of undiscovered porphyry copper resources in the Andes Mountains of South American: Society of Economic Geology Newsletter, no. 71, pp. 1, 8–13.

Cunningham, C.G., Singer, D.A., Zappettini, E.O., Vivallo, S.W., Celada, C.M., Quispe, J., Briskey, J.A., Sutphin, D.M., Gajardo, M.M., Diaz, A., Portigiliati, C., Berger, V.I., Carrasco, R., and Schultz, K.J., 2008, Quantitative mineral resource assessment of copper, molybdenum, silver, and gold in undiscovered porphyry copper deposits in the Andes Mountains of South America: U.S. Geological Survey Open File Report 2008–1253. Available at: http://pubs.usgs.gov/of/2008/1253/.

Dallal, G.E., and Wilkinson, L., 1986, An analytical approximation to the distribution of Lilliefors' test statistic for normality: American Statistician, v. 40, no. 4, pp. 294–295.

Davis, G.A., and Samis, M., 2006, Using real options to value and manage exploration, *in* Doggett, M.D., and Parry, J.R., eds., Wealth creation in the mineral industry—integrating science, business, and education: Society of Economic Geologists Special Publication 12, pp. 273–294.

De Veaux, R.D., and Velleman, P.F., 2008, Math is music; statistics is literature (or why are there no six-year-old novelists?): AMSTAT NEWS (American Statistical Association), September, pp. 54–58.

DeYoung, J.H., Jr., 1981, The Lasky cumulative tonnage-grade relationship—a re-examination: Economic Geology, v. 76, no. 5, pp. 1067–1080.

DeYoung, J.H., Jr., and Menzie, W.D., 1999, The changing uses of minerals information—a government perspective, *in* Otto, J., and Kim, H.-S., eds., Proceedings of the Workshop on the Sustainable Development of Non-renewable Resources, Towards the 21st Century: New York, 15–16 October 1998: United Nations Development Programme, New York, pp. 111–127.

DeYoung, J.H., Jr., and Singer, D.A., 1981, Physical factors that could restrict mineral supply, *in* Skinner, B.J., ed., Economic Geology, 75th anniversary volume, pp. 939–954.

Drew, L.J., 1975, Linkage effects between deposit discovery and post discovery exploratory drilling: U.S. Geological Survey Journal of Research, v. 3, no. 2, pp. 169–179.

Drew, L.J., 1990, Oil and gas forecasting—reflections of a petroleum geologist: Oxford University Press, New York.

Drew, L.J., 1997, Undiscovered petroleum and mineral resources—assessment and controversy: Plenum Press, New York.

Drew, L.J., Bliss, J.D., Bowen, R.W., Bridges, N.J., Cox, D.P., DeYoung, J.H., Houghton, J.C., Ludington, S., Menzie, W.D., Page, N.J, Root, D.H., and Singer, D.A., 1986, Quantitative estimation of undiscovered mineral resources—the case study of U.S. Forest Service wilderness tracts in the Pacific mountain system: Economic Geology, v. 81, no. 1, pp. 80–88.

Drew, L.J., and Menzie, W.D., 1993, Is there a metric for mineral deposit occurrence probabilities?: Nonrenewable Resources, v. 2, n. 2, pp. 92–105.

Duval, J.S., 2004, EMINERS—an economic mineral resource simulator: U.S. Geological Survey Open File Report 04–1344. Available at: http://pubs.usgs.gov/of/2004/1344.

Eberlein, D.G., and Menzie, W.D., 1978, Maps and tables describing areas of mineral resource potential of Central Alaska: U.S. Geological Survey Open File Report78–1-D.

Eckstrand, O.R., ed., 1984, Canadian mineral deposit types, a geological synopsis: Geological Survey of Canada, Economic Geology Report 36.

Eckstrand, O.R., Sinclair, W.D., and Thorpe, R.I., eds., 1995, Geology of Canadian mineral deposit types: Geological Survey of Canada, Geology of Canada, no. 8, and Geological Society of America's Geology of North America, v. P-1.

Edelstein, D.L., 2002, Copper, *in* U.S. Geological Survey mineral commodity summaries 2002, pp. 54–55. Available at: http://minerals.usgs.gov/minerals/pubs/mcs/.

Erickson, R.L., 1973, Crustal abundance of elements, and mineral reserves and resources, *in* Brobst, D.A., and Pratt, W.P., eds., United States mineral resources: U.S. Geological Survey Professional Paper 820, pp. 21–25.

Fahrni, K.C., McCauley, T.N., and Preto, V.A., 1976, Copper Mountain and Ingerbelle, *in* Sutherland Brown, A., ed., Porphyry deposits of the Canadian Cordillera: Canadian Institute Mining and Metallurgy Special Paper 15, pp. 368–375.

Gilbert, R.O., 1987, Statistical methods for environmental pollution monitoring: Van Nostrand Reinhold, New York.

Griffiths, J.C., and Ondrick, C.W., 1970, Structure by sampling in the geosciences, *in* Patil, G.P., ed., Random counts in physical science, geoscience, and business: Penn State Statistics Series, Pennsylvania State University Press, University Park, Pa., pp. 31–55.

Grybeck, D., and DeYoung, J.H., Jr., 1978, Maps and tables describing areas of mineral resource potential of the Brooks Range, Alaska: U.S. Geological Survey Open File Report 78–1-B.

Gustafson, L.B., and Hunt, J.P., 1975, The porphyry copper deposit at El Salvador, Chile: Economic Geology, v. 70, pp. 857–912.

Gunther, T.M., 1992, Quantitative assessment of future development of copper/silver resources in the Kootenai National Forest, Idaho/Montana: part II—estimation of the copper and silver endowments: Nonrenewable Resources, v. 1, n. 4, pp. 267–280.

Harris, D.P., 1965, An application of multivariate statistical analysis to mineral exploration: Ph.D. dissertation, Pennsylvania State University.

Harris, D.P., 1984, Mineral resources appraisal—mineral endowment, resources, and potential supply—concepts, methods, and cases: Oxford University Press, Oxford, U.K.

Harris, D., and Rieber, M., 1993, Evaluation of the United States Geological Survey's three-step assessment methodology: U.S. Geological Survey Open File Report 93–258-A.

Hedger, D., 2008, Quantification of economic risk in mineral exploration: a case study on exploration in the Yarrol region, Queensland, Australia: Ph.D. dissertation, University of Queensland.

Hodges, C.A., Cox, D.P,. Singer, D.A., Case, J.E., Berger, B.R., and Albers, J. P. 1984, U.S. Geological Survey—INGEOMINAS mineral resource assessment of Colombia: U.S. Geological Survey Open File Report 84–345.

Hoover, D.B., Heran, W.D., and Hill, P.L., eds., 1992, The geophysical expression of selected mineral deposit models: U.S. Geological Survey Open File Report 92–557.

Hudson, T., and DeYoung, J.H., Jr., 1978, Maps and tables describing areas of mineral resource potential, Seward Peninsula, Alaska: U.S. Geological Survey Open File Report 78–1-C.

Jachens, R.C., Moring, B.C., and Schruben, P.G., 1996, Thickness of Cenozoic deposits and the isostatic residual gravity over basement in Nevada, *in* Singer, D.A., ed., An analysis of Nevada's metal-bearing mineral resources: Nevada Bureau of Mines and Geology Open File Report 96–2, chap. 2.

King, Clarence, 1880, First annual report of the United States Geological Survey to the secretary of the interior: Annual Report 1, United States Geological Survey.

Kirkham, R.V., Sinclair, R.V., Thorpe, W.D., and Duke, J.M., eds., 1993, Mineral deposit modeling: Geological Association of Canada Special Paper 40.

Kording, K.P., and Wolpert, D.W., 2004, Bayesian integration in sensorimotor learning: Nature, v. 427, no. 2, pp. 244–247.

Kouda, R., and Singer, D.A., 1992, A resource assessment of kuroko deposits in the Akita region of Japan: 29th International Geologic Congress, Kyoto, Abstracts, v. 1, p. 83.

Langton, J.M., and Williams, S.A., 1982, Structural petrological, and mineralogical controls for the Dos Pobres ore body, in Titley, S.R., ed., Advances in geology of the porphyry copper deposits, southwestern North America: University of Arizona Press, Tucson, pp. 335–352.

Lanier, George, John, Bard, C.E., Caddy, S.W., and Wilson J.C., 1978, General geology of the Bingham mine, Bingham Canyon, Utah: Economic Geology, v. 73, pp. 1228–1241.

Lasky, S.G., 1950, How tonnage and grade relations help predict ore reserves: Engineering and Mining Journal, v. 151, no. 4, pp. 81–85.

Lefebure, D.V., and Hoy, T., eds., 1996, Metallic deposits, v. 2 of Selected British Columbia mineral deposit profiles: British Columbia Ministry of Employment and Investment Open File 1996–13.

Lefebure, D.V., and Ray, G.E., 1995, Metallic and coal, v. 1 of Selected British Columbia mineral deposit profiles: British Columbia Ministry of Energy, Mines and Petroleum Resources Open File 1995–20.

Lisitsin, V., Olshina, A., Moore, D.H., and Willman, C.E., 2007, Assessment of undiscovered mesozonal orogenic gold endowment under cover in the northern part of the Bendigo Zone: GeoScience Victoria Gold Undercover Report 2, Department of Primary Industries. Available at: http://www.dpi.vic.gov.au/minpet/store.

Lisitsin, V., Olshina, A., Moore, D.H., and Willman, C.E., 2009, Assessment of undiscovered mesozonal orogenic gold endowment under cover in the northern part of the Stawell Zone (Victoria): Gold Undercover Report 13, Department of Primary Industries. Available at: http://www.dpi.vic.gov.au/minpet/store.

Long, K.R., 2009, A test and re-estimation of Taylor's empirical capacity-reserve relationship: Natural Resources Research, v. 18, no. 1, pp. 57–63.

Long, K.R., and Singer, D.A., 2001, A simplified economic filter for open-pit mining and heap-leach recovery of copper in the United States: U.S. Geological Survey Open File Report 01–218. Available: http://geopubs.wr.usgs.gov/open-file/of01–018/

Lowell, J.D., and Guilbert, J.M., 1970, Lateral and vertical alteration-mineralization zoning in porphyry copper deposits: Economic Geology, v. 65, pp. 63–408.

Luddington, S., Hammarstrom, J., Robinson, G., Gray, F., Vélez-López, J., Salinas-Prieto, J.C., Alba-Infante, H., Cendejas-Cruz, F., Espinosa, E., Valencia-Morena, M., Rodríguez-Castañeda, J.L., and Zürcher, L., 2008, Mineral resource assessment using digital geospatial data-porphyry copper deposits in Mexico: 33rd International Geologic Congress, 6–14 August 2008, Oslo, Norway.

Ludington, S., and Cox, D.P., eds., 1996, Data base for a national mineral resource assessment of undiscovered deposits of gold, silver, copper, lead, and zinc in the conterminous United States: United States Geological Survey, Open File Report 96–96.

Mackenzie, B.W., 1994, Evaluating and controlling geological risk, *in* Jones, H., chair, Managing Risk; 4th Large Open Pit Mining Conference Proceedings: Australasian Institute of Mining and Metallurgy, 5–9 September 1994, Perth, Australia, pp. 47–55.

MacKevett, E.M., Jr., Singer, D.A., and Holloway, C.D., 1978, Maps and tables describing metalliferous mineral resource potential of Southern Alaska: U.S. Geological Survey Open File Report 78–1-E.

Marcus, S.M., Frieders, T., Juilland, J., and Pavek, G., 1986, Program review and status report of mineral surveys by the Geological Survey and the Bureau of Mines of certain lands managed by the Bureau of Land Management: U.S. Geological Survey Open File Report 86–165.

Matheron, G., 1959, Remarques sur la loi de Lasky: Chronique des Mines d'Outre-Mer et de la Recherche Miniere, 27e annee, no. 282, pp. 463–465.

McCammon, R.B., Menzie, W.D., Mast, R.F., and Carter, M.D., 1991, Quantitative assessments of the energy and mineral resources within eighteen wilderness study areas in the States of Colorado, Nevada, Oregon, and Utah: U.S. Geological Survey Open File Report 91–384.

McKelvey, V.E., 1960, Relations of reserves of the elements to their crustal abundance: American Journal of Science, v. 258-A (Bradley volume), pp. 234–241.

McKelvey, V.E., 1973, Mineral resource estimates and public policy, *in* Brobst, D.A., and Pratt, W.P., eds., United States mineral resources: U.S. Geological Survey Professional Paper 820, pp. 9–19.

Menzie, W.D., DeYoung, J.H., Jr., and Steblez, W.G., 2001, Some implications of changing patterns of mineral consumption, *in* World Mining Congress XVIII at MINExpo INTERNATIONAL 2000, Las Vegas, Nev., 9–12 October 2000: U.S. Geological Survey Open File Report 03–382. Available at: http://pubs.er.usgs.gov/usgspubs/ofr/ofr03382.

Menzie, W.D., and Singer, D.A., 1993, Grade and tonnage model for porphyry Cu deposits in British Columbia, Canada, and Alaska, United States: U.S. Geological Survey Open File Report 93–275.

Menzie, W.D., Singer, D.A., and DeYoung, J.D., Jr., 2005, Physical resource limitations, *in* Simpson, David, Toman, M.A., and Ayres, R.U. eds., Scarcity and growth revisited—natural resources and the environment in the new millennium: Resources for the Future Press, Washington, D.C., pp. 33–53.

Menzie, W.D., Singer, D.A., Karangan, N., and Tresnadi, H., 1997, The Hila Prospect: a recently discovered copper occurrence on Ambon Island, Republic of Indonesia: U.S. Geological Survey Open File Report 97–86.

Meyer, M.A., and Booker, J.M., 2001, Eliciting and analyzing expert judgment: a practical guide: American Statistical Association and Society for Industrial and Applied Mathematics, Philadelphia, Pa.

Mining Journal, 1998, Poverty and risk: v. 330, no. 8471, pp. 208–209.

Moore, D.P., and Drew, L.J., 1979, A minerals-exploration simulation model: Proceedings of the Council of Economics, v. 108, pp. 77–83.

Morris, H.T., 1986, Descriptive model of polymetallic replacement deposits, *in* Cox, D.P., and Singer, D.A., eds., 1986, Mineral deposit models: U.S. Geological Survey Bulletin 1693, pp. 99–100.

Mosier, D.L., Berger, V.I., and Singer, D.A., 2009, Volcanogenic massive sulfide deposits of the world—database and grade and tonnage models: U.S. Geological Survey Open File Report 2009–1034. Available at: http://pubs.usgs.gov/of/2009/1034/.

Mosier, D.L., and Page, N.J, 1988, Descriptive and grade–tonnage models of volcanogenic manganese deposits in oceanic environments—a modification: U.S. Geological Survey Bulletin 1811.

Mosier, D.L., Singer, D.A., and Berger, V.I., 2007, Volcanogenic massive sulfide deposit density; U.S. Geological Survey Scientific Investigations Report 2007–5082. Available at: http://pubs.usgs.gov/sir/2007/5082/.

Mosier, D.L., Singer, D.A., and Salem, B.B., 1983, Geologic and grade-tonnage information on volcanic-hosted copper-zinc-lead massive sulfide deposits: U.S. Geological Survey Open File Report 83–89.

Murphy, A.H., and Winkler, R.L., 1984, Probability forecasting in meteorology: Journal American Statistical Association, v. 79, no. 387, pp. 489–500.

Ohmoto, H., and Takahashi, T., 1983, Submarine calderas and kuroko genesis, *in* Omoto, H., and Skinner, B.J., eds., The kuroko and related volcanogenic massive sulfide deposits: Economic Geology Monograph 5, pp. 38–54.

Orris, G.J., and Bliss, J.D., 1985, Geologic and grade-volume data on 330 gold placer deposits: U.S. Geological Survey Open File Report 85–213.

Orris, G.J., and Bliss, J.D., eds., 1991, Some industrial mineral deposit models–Descriptive deposit models: U.S. Geological Survey Open File Report 91–11A.

Orris, G.J., and Bliss, J.D., eds., 1992, Industrial mineral deposit models—grade and tonnage of models: U.S. Geological Survey Open File Report 92–437.

Penney, S.R., Allen, R.M., Harrisson, S., Lees, T.C., Murphy, F.C., Norman, A.R., and Roberts, P.A., 2004, A global-scale exploration risk analysis technique to determine the best mineral belts for exploration: Transactions of the Institution of Mining and Metallurgy, Section B, Applied Earth Science, v. 113, pp. B183–B196.

Porter, K.E., and Edelstein, D.L., 2008, Copper statistics: historical statistics for mineral and material commodities in the United States: U.S. Geological Survey Data Series 140. Available at: http://pubs.usgs.gov/ds/2005/140/.

Porwal, A., 2007, Base-metal prospectivity mapping of the Aravalli Province—a spatial mathematical approach: Geological Society of Australia Newsletter, December, no. 4, pp. 3–19.

Pride, D.E., Cox, C.A., Moody, S.V., Conelea, R.R., and Rosen, M.A., 1990, Investigation of mineralization in the South Shetland Islands, Gerlache Strait, and Anvers Island, Northern Antarctic Peninsula, *in* Splettstoesser, J.F., and Dreschhoff, G.A.M., eds., Mineral resources potential of Antarctica: American Geophysical Union, Washington D.C., v. 51, pp. 69–94.

Raines, G.L., and Mihalasky, M.J., 2002, A reconnaissance method for delineation of tracts for regional-scale mineral resource assessment based on geologic-map data: Natural Resources Research, v. 11, no. 4, pp. 241–248.

Rasumovsky, N.K., 1940, Distribution of metal values in ore deposits: Acad. Sci. Comptes Rendus (Doklady) U.R.S.S., v. 28, pp. 814–816.

Reed, B.L., Menzie, W.D., McDermott, M., Root, D.H., Scott, W., and Drew, L.J., 1989, Undiscovered lode tin resources of the Seward Peninsula, Alaska: Economic Geology, v. 84, no. 7, pp. 1936–1947.

Richter, D.H., Singer, D.A., and Cox, D.P., 1975, Mineral resource map of the Nabesna Quadrangle, Alaska: U.S. Geological Survey Miscellaneous Field Studies Map MF-655K.

Roberts, R.G., and Sheahan, P.A., eds., 1988, Ore deposit models: Geoscience Canada Reprint Series 3.

Rogers, M.C., Thurston, P.C., Fyon, J.A., Kelly, R.I., and Breaks, R.W., 1995, Descriptive mineral deposit models of metallic and industrial deposit types and related mineral potential assessment criteria: Ontario Geological Survey Open File Report 5916.

Rongfu, Pei, ed., 1995, Mineral deposits models of China: Geological Publishing House, Beijing.

Root, D.H., Menzie, W.D., and Scott, W.A., 1992, Computer Monte Carlo simulation in quantitative resource estimation: Nonrenewable Resources, v. 1, no. 2, pp. 125–138.

Rose, P.R., 1992, Risk behavior in petroleum exploration, *in* Steinmetz, Richard, ed., The business of petroleum exploration: American Association of Petroleum Geologists, Tulsa, Okla., pp. 95–103.

Rowley, P.D., Farrar, E., Carrara, P.E., Vennum, W.R., and Kellogg, K.S., 1988, Porphyry copper deposits and potassium–argon ages of plutonic rocks of the Orville Coast and Eastern Ellsworth Land, Antarctica: U.S. Geological Survey Professional Paper 1351C, pp. 35–49.

Rowley, P.D., Ford, A.B., Williams, P.L., and Pride, D.E., 1975, Metallogenic provinces of Antarctica, *in* R.L. Oliver, P.R. James, and J.B. Jago, eds., Antarctic Earth Science: Canberra, Australian Academy of Science, pp. 414–419.

Rowley, P.D., Williams, P.L., and Pride, D.E., 1991, Metallic and non-metallic mineral resources of Antarctica, *in* Tingley, R.J., ed., The geology of Antarctica: Clarendon Press, Oxford, U.K., pp. 617–651.

Rowley, P.D., Williams, P.L., Schmidt, D.L., Reynolds, R.L., Ford, A.B., Clark, A.B., Farrar, E., and McBride, S.L., 1983, Copper mineralization along the Lassiter Coast of the Antarctic Peninsula: Australian Academy of Science, Canberra, pp. 414–419.

Scheibner, E., 1998, Geology of New South Wales—synthesis, v. 2, geological evolution: Geological Survey of New South Wales, Memoir Geology 13(2).

Schruben, P., 2002, Assessment of undiscovered deposits of gold, silver, copper, lead, and zinc in the United States: a portable document (PDF) recompilation of USGS OFR 96–96 and Circular 1178 (1998): U.S. Geological Survey Open File Report 02–198.

Schruben, P.G., Wynn, J.C., Gray, F., Cox, D.P., Stewart, J.H., and Brooks, W.E., 1997, geology and resource assessment of the Venezuelan Guayana Shield: Digital Data Series 46. Available at: http://pubs.usgs.gov/dds/dds46/.

Schulz, K.J., and Briskey, J.A., 2007, U.S. Geological Survey national mineral resource assessment—an estimate of undiscovered deposits of gold, silver, copper, lead, and zinc in the United States, *in* Briskey, J.A., and Schulz, K.J., eds., Proceedings for a Workshop on Deposit Modeling, Mineral Resources Assessment, and Their Role in Sustainable Development: U.S. Geological Survey Circular 1294, pp. 131–135. Available at: http://pubs.usgs.gov/circ/2007/1294/.

Scott, M., 2000, Valuing Australian state geological surveys: quantitative analysis for strategic planning: Ph.D. dissertation, University of Queensland.

Scott, M., and Dimitrakopoulos, R., 2001, Quantitative analysis of mineral resources for strategic planning—implications for Australian geological surveys: Natural Resources Research, v. 10, no. 3, pp. 159–177.

Shapiro, S.S., Wilk, M.B., and Chen, H.J., 1968, A comparative study of various tests for normality: Journal of the American Statistical Association, v. 63, no. 324, pp. 1343–1372.

Sheahan, P.A., and Cherry, M.E., eds., 1993, Ore deposit models, v. 2: Geoscience Canada Reprint Series 6.

Shulman, M.J., Burger, H., Kliem, C., and Skala, W., 1992, Review of resource analysis and related studies: Working Group, Mathematical Geology, Freie Universität Berlin: Nonrenewable Resources, v. 1, no. 1, pp. 97–106.

Sillitoe, R.H., 1979, Some thoughts on gold-rich porphyry copper deposits: Mineralium Deposita, v. 14, no. 2, pp. 161–174.

Sillitoe, R.H., 1995, Exploration and discovery of base-and precious metal deposits in the circum-Pacific region during the last 25 years: Resource Geology Special Issue No. 19.

Singer, D.A., 1975, Mineral resource models and the Alaskan Mineral Resource Assessment Program, *in* Vogely, W.A., ed., Mineral materials modeling: a state-of-the-art review: Johns Hopkins University Press, Baltimore, pp. 370–382.

Singer, D.A., 1986, Descriptive model of kuroko massive sulfide, *in* Cox, D.P., and Singer, D.A., eds., Mineral deposit models: U.S. Geological Survey Bulletin 1693, pp. 189–190.

Singer, D.A., 1992, Grade and tonnage model of Sierran kuroko deposits, *in* Bliss, J.D., ed., Developments in deposit modeling: U.S. Geological Survey Bulletin 2004, pp. 29–32.

Singer, D.A., 1993a, Basic concepts in three-part quantitative assessments of undiscovered mineral resources: Nonrenewable Resources, v. 2, no. 2, pp. 69–81.

Singer, D.A., 1993b, Development of grade and tonnage models for different deposit types, *in* Kirkham, R.V., Sinclair, W.D., Thorpe, R.I., and Duke, J.M., eds., Mineral deposit modeling: Geological Association Canada Special Paper 40, pp. 21–30.

Singer, D.A., 1994a, Conditional estimates of the number of podiform chromite deposits: Nonrenewable Resources, v. 3, no. 3, pp. 200–204.

Singer, D.A., 1994b, The relationship of estimated number of undiscovered deposits to grade and tonnage models in three-part mineral resource assessments, *in* 1994 International Association Mathematical Geology Annual Conference, papers and extended abstracts, 3–5 October 1994, Mount Tremblant, Quebec, pp. 325–326.

Singer, D.A., 1995, World-class base and precious metal deposits—a quantitative analysis: Economic Geology, v. 90, no. 1, pp. 88–104.

Singer, D.A., 1996, Grade and tonnage models, *in* Singer, D.A., ed., An analysis of Nevada's metal-bearing mineral resources: Nevada Bureau of Mines and Geology Open File Report 96–2, chap. 11, pp. 11.1–11.18. Available at: ftp://ftp.nbmg.unr.edu/NBMG/nbmg_ofr96_2/.

Singer, D.A., 2001, Some suggested future directions of quantitative resource assessments: Journal of China University of Geosciences, v. 12, no. 1, pp. 40–44 (in Chinese in v. 26, no. 2, pp. 152–156).

Singer, D.A., 2005, Planning exploration under cover, *in* Proceedings of 20th World Mining Congress and Expo 2005, 7–11 November 2005, Tehran, Iran, pp. 353–356.

Singer, D.A., 2006, Typing mineral deposits using their associated rocks and grades and tonnages in a probabilistic neural network: Mathematical Geology, v. 38, no. 4, pp. 465–475.

Singer, D.A., 2007, Estimating amounts of undiscovered mineral resources, *in* Briskey, J.A., and Schulz, K.J., eds., Proceedings for a Workshop on Deposit Modeling, Mineral Resources Assessment, and Their Role in Sustainable Development: U.S. Geological Survey Circular 1294, pp. 79–84. Available at: http://pubs.usgs.gov/circ/2007/1294/.

Singer, D.A., 2008, Mineral deposit densities for estimating mineral resources: Mathematical Geosciences, v. 40, no. 1, pp. 33–46.

Singer, D.A., Berger, V.I., Menzie, W.D., and Berger, B.B., 2005, Porphyry copper density: Economic Geology, v. 100, no. 3, pp. 491–514.

Singer, D.A., Berger, V.I., and Moring, B.C., 2005, Porphyry copper deposits of the world: database, map, and grade and tonnage models: U.S. Geological Survey Open File Report 2005–1060. Available at: http://pubs.usgs.gov/of/2005/1060/.

Singer, D.A., Berger, V.I., and Moring, B.C., 2008, Porphyry copper deposits of the world: database, map, and grade and tonnage models, 2008: U.S. Geological Survey Open File Report 2008–1155. Available at: http://pubs.usgs.gov/of/2008/1155/.

Singer, D.A., Cox, D.P., and Drew, L.J., 1975, Grade and tonnage relationships among copper deposits: U.S. Geological Survey Professional Paper 907A, pp. A1–A11.

Singer, D.A., and DeYoung, J.H., Jr., 1980, What can grade and tonnage relations really tell us? *in* Guillemin, C., and Lagny, P., eds., Ressources minerale—mineral resources: Bureau de Recherches Géologiques et Miniéres Memoires 106, pp. 91–101.

Singer, D.A., and Kouda, R., 1997a, Classification of mineral deposits into types using mineralogy with a probabilistic neural network: Nonrenewable Resources, v. 6, no. 1, pp. 27–32.

Singer, D.A., and Kouda, R., 1997b, Use of a neural network to integrate geoscience information in the classification of mineral deposits and occurrences, *in* Gubins, A.G., ed., Proceedings of Exploration 97: Fourth Decennial International Conference on Mineral Exploration, pp. 127–134.

Singer, D.A., and Kouda, R., 1999a, Examining risk in mineral exploration: Natural Resources Research, v. 8, no. 2, pp. 111–122.

Singer, D.A., and Kouda, R., 1999b, A comparison of the weights of evidence method and probabilistic neural networks: Natural Resources Research, v. 8, no. 4, pp. 287–298.

Singer, D.A., and Kouda, R., 2003, Typing mineral deposits using their grades and tonnages in an artificial neural network: Natural Resources Research, v. 12, no. 3, pp. 201–208.

Singer, D.A., and Kouda, R., 2006, Reducing risk in exploration under cover, *in* Koike, K., Notosiswoyo, S., Kouda, R., and Sulistianto, B., eds., Proceedings of 9th International Symposium on Mineral Exploration (ISME IX): Toward new frontiers for resource exploration and sustainable development: Penerbit ITB, 19–21 September 2006, Bandung, Indonesia, pp. 16–21.

Singer, D.A., and Kouda, R., 2008, Probabilistic estimates of number of mineral deposits using deposit densities, *in* Abstract CD-ROM, 33rd International Geologic Congress, 6–14 August 2008, Oslo, Norway.

Singer, D.A., and MacKevett, E.M., 1977, Mineral resources map of the McCarthy Quadrangle, Alaska: U.S. Geological Survey Miscellaneous Field Studies Map MF 773-C.

Singer, D.A., and Menzie, W.D., 2005, Statistical guides to estimating the number of undiscovered mineral deposits: an example with porphyry copper deposits, *in* Cheng, Q., and Bonham-Carter, G., eds., Proceedings of IAMG—the annual conference of the International Association for Mathematical Geology: Geomatics Research Laboratory, York University, Toronto, Canada, pp. 1028–1033.

Singer, D.A., and Menzie, W.D., 2008, Map scale effects on estimating the number of undiscovered mineral deposits: Natural Resources Research, v. 17, no. 2, pp. 79–86. (Also available *in* Bonham-Carter, G., and Cheng, Q., eds, Progress in geomathematics: Springer-Verlag, Berlin, pp. 271–283.)

Singer, D.A., Menzie, W.D., DeYoung, J.H., Jr., Sander, M., and Lott, A., 1980, Grade and tonnage data used to construct models for the Regional Alaskan Mineral Resource Assessment Program: U.S. Geological Survey Open File Report 80–799.

Singer, D.A., Menzie, W.D., and Long, K.R., 1998, A simplified economic filter for open-pit gold-silver mining in the United States: U.S. Geological Survey Open File Report 98–207. Available: http://geopubs.wr.usgs.gov/open-file/of98–207/.

Singer, D.A., Menzie, W.D., and Long, K.R., 2000, A simplified economic filter for underground mining of massive sulfide deposits: U. S. Geological Survey Open File Report 00–349.

Singer, D.A., Menzie, W.D., Sutphin, D., Mosier, D.L., and Bliss, J.D., 2001, Mineral deposit density—an update, *in* Schulz, K.J., ed., Contributions to global mineral resource assessment research: U.S. Geological Survey Professional Paper 1640-A, pp. A1–A13. Available at: http://pubs.usgs.gov/prof/p1640a/.

Singer, D.A., and Mosier, D.L., 1981a, Review of regional mineral resource assessment methods: Economic Geology, v. 76, no. 5, pp. 1006–1015.

Singer, D.A., and Mosier, D.L., 1981b, The relation between exploration economics and the characteristics of mineral deposits, *in* Ramsey, J.B., ed., The economics of exploration for energy resources: JAI Press, New York, pp. 313–326.

Singer, D.A., and Mosier, D.L., eds., 1983a, Mineral deposit grade–tonnage models: U.S. Geological Survey Open File Report 83–623.

Singer, D.A., and Mosier, D.L., eds., 1983b, Mineral deposit grade–tonnage models—II: U.S. Geological Survey Open File Report 83–902.

Singer, D.A., Mosier, D.L., and Cox, D.P., 1986, Grade-tonnage model of porphyry copper, *in* Cox, D.P., and Singer, D.A., eds., Mineral deposit models: U.S. Geological Survey Bulletin 1693, pp. 77–81.

Singer, D.A., Mosier, D.L., and Menzie, W.D., 1993, Digital grade and tonnage data for 50 types of mineral deposits—Macintosh version: U.S. Geological Survey Open File Report 93–280.

Singer, D.A., and Ovenshine, A.T., 1979, Assessing metallic mineral resources in Alaska: American Scientist, v. 67, no. 5, pp. 582–589. (Republished *in* Skinner, B.J., ed., 1980, Earth's Energy and Mineral Resources, William Kaufman, Los Altos, pp. 170–177).

Singer, D.A., Page, N.J, Smith, J.G., Blakely, R.J., and Johnson, M.G., 1983, Mineral resource assessment of the Medford 1° × 2° Quadrangle, Oregon-California: U.S. Geological Survey Miscellaneous Field Studies Map MF-1383-E.

Singer, D.A., Waller, N., Mosier, D.L., and Bliss, J.D., 1997, Digital mineralogy data for 55 types of mineral deposits: Macintosh version: U.S. Geological Survey Open File Report 97–160.

Skinner, B.J., 1976, A second iron age ahead?: American Scientist, v. 64, no. 3, pp. 258–269.

Slichter, L.B., 1960, The need of a new philosophy of prospecting: Mining Engineering, June, pp. 570–575.

Smith, R.C., 1992, PREVAL—prefeasibility software program for evaluating mineral deposits: U.S. Bureau of Mines Information Circular 9307.

Spanski, G.T., 1992, Quantitative assessment of future development of copper/silver resources in the Kootenai National Forest, Idaho/Montana: part I—estimation of the copper and silver endowments: Nonrenewable Resources, v. 1, no. 2, pp. 163–183.

Stanley, M.C., 1992, Statistical trends and discoverability modeling of gold deposits in the Abitibi greenstone belt, Ontario, *in* Kim, Y.C., ed., Proceedings of the 23rd International Symposium on Applications of Computers and Operations Research in the Mineral Industry APCOM'92, 7–11 April 1992, Tucson, AZ: Society for Mining, Metallurgy, and Exploration/ American Institute of Mining, Metallurgical, and Petroleum Engineers, New York, pp. 17–28.

Stern, H., 1991, On the probability of winning a football game: American Statistician, v. 45, no. 3, pp. 179–183.

Stuart, A., and Ord, J.K., 1991, Kendall's advanced theory of statistics, v. 2, 5th ed.: Oxford University Press, New York.

Taylor, H.K., 1978, Mine valuation and feasibility studies, *in* Hoskins, J.R., and Green, W.R., eds., Mineral industry costs, 2nd. ed.: Northwest Mining Association, Spokane, Wash., pp. 1–17.

Taylor, H.K., 1985, Cutoff grades—some further reflections: Transactions of the Institution of Mining and Metallurgy, Section A, Mining Industry, v. 94, pp. A204–A216.

Taylor, H.K., 1986, Rates of working mines; a simple rule of thumb: Transactions of the Institution of Mining and Metallurgy, Section A, Mining Industry, v. 95, pp. 203–204.

Theodore, T.G., Orris, G.A., Hammarstrom, J.M., and Bliss, J.D. 1991, Gold–bearing skarns: U.S. Geological Survey Bulletin 1930.

Titley, S.R., 1982, The style and progress of mineralization and alteration and alteration in porphyry copper systems, in Titley, S.R., ed., Advances in geology of the porphyry copper deposits: Tucson, University of Arizona Press, pp. 93–116.

U.S. Geological Survey National Mineral Resource Assessment Team, 1998, 1998 Assessment of undiscovered deposits of gold, silver, copper, lead, and zinc in the United States: U.S. Geological Survey Circular 1178.

U.S. Geological Survey and Servicio Geologico de Bolivia, 1992, Geology and mineral resources of the Altiplano and Cordillera Occidental, Bolivia—with a section on application of economic evaluations to deposit models, by Bleiwas, D.I., and Christiansen, R.G.: U.S. Geological Survey Bulletin 1975, 218–219.

Wellmer, F.-W., Dalheimer, M., and Wagner, M., 2008, Economic evaluations in exploration, 2nd ed.: Springer-Verlag, Berlin.

Wernick, I.K., Herman, R., Govind, S., and Ausubel, J.H., 1996, Materialization and dematerialization—measures and trends: Daedalus (Journal of the American Academy of Arts and Sciences), v. 125, no. 3, pp. 171–198.

Wilford, J., 2000, Thematic mapping and three-dimensional modeling of the regolith for mineral and environmental assessment, *in* Abstracts CD, 31st International Geologic Congress, Rio de Janeiro, Brazil.

Wilkinson, B.H., and Kesler, S.E., 2007, Tectonism and exhumation in convergent margin orogens: insights from ore deposits: Journal of Geology, v. 115, pp. 611–627.

Wilson, F.H., White, W.H., Detterman, R.L., and Case, J.E., 1996, Maps showing the resource assessment of the Port Moller, Stepovak Bay, and Simeonof Island Quadrangles, Alaska Peninsula: U.S. Geological Survey MF 2155-F.

Index

The manufacturer's authorised representative in the EU for product safety is Oxford University Press España S.A. of El Parque Empresarial San Fernando de Henares, Avenida de Castilla, 2 - 28830 Madrid (www.oup.es/en or product.safety@oup.com). OUP España S.A. also acts as importer into Spain of products made by the manufacturer.
Printed and bound by CPI Group (UK) Ltd, Croydon, CR0 4YY
12/09/2026
02215931-0001